THE CIRCUIT RIDER OF
MISSISSIPPI:
— An Indian Immigrant's Story in the Post-JimCrow South —

Jerry F. Miller, Yashas Hariprasad, Ranvir Iyengar

The Circuit Rider
of Mississippi

Jerry F. Miller
Yashas Hariprasad
Ranvir Iyengar

Dedication

This book is dedicated to Nelson Mandela, whose unwavering courage and vision for justice continue to inspire generations.

His legacy reminds us that perseverance and wisdom can transform the world.

Foreword

Dr. S.S. Iyengar's journey is deeply intertwined with the historical legacy of the Jim Crow South, a time when the doctrine of "separate but equal" defined daily life, enforcing strict racial segregation and systemic inequality. Mississippi, like much of the South, was marked by deep-rooted racial divides that limited opportunities for people of color, reinforcing economic and educational disparities. In this environment, where the rights of minorities were legally curtailed and racial prejudice was institutionalized, Dr. Iyengar's path was profoundly challenging.

Navigating the segregated world of the Jim Crow South required more than academic prowess; it demanded a resilience and quiet fortitude that few could summon. Dr. Iyengar confronted social barriers that extended beyond classrooms and into every facet of life, from limited access to educational resources to the often-unspoken rules governing interactions in academic and professional circles. These were not mere inconveniences but barriers intentionally designed to deter and dishearten individuals like him.

Yet, as the book reveals, Dr. Iyengar's resolve only grew in the face of these systemic challenges. His experiences during this period shaped his belief in the transformative power of education and strengthened his commitment to creating opportunities for others. It is this resilience and dedication that shines throughout *The Circuit Rider of Mississippi*, illustrating how, even within the constraints of a society structured by segregation, he found ways to bridge divides, excel, and ultimately impact countless lives across the globe.

The story of Dr. Iyengar's life is a testament to the triumph of the human spirit over the legacy of "separate but not equal." It demonstrates that, despite the profound obstacles imposed by Jim Crow laws, true progress is driven by vision, strength, and a refusal to be bound by society's limitations. Through his achievements, he broke barriers for himself and forged a path for others to follow.

Having read early drafts of this book several years ago, I am deeply moved by the final narrative, which so vividly captures the essence of Dr. Iyengar's life—from his modest beginnings in the deep South to his profound influence on the global academic stage. Although the book was originally slated for publication in 2019, unforeseen circumstances delayed its release.

I am proud to have witnessed Dr. Iyengar's journey firsthand and to introduce this compelling account of his life. The Circuit Rider of Mississippi is far more than just a biography— it is a powerful reminder of the resilience of the human spirit and the limitless potential that resides within us all.

– Dr. R.L Kashyap

Padma Shri Awardee
Emeritus Professor
Purdue University

May, 2021

Preface

Diving beyond the ordinary immigrant narrative, "The Circuit Rider of Mississippi - An Indian Immigrant's Story in the Post-Jim Crow South" intricately weaves the turbulent technological shifts of a young Indian immigrant in the USA. This narrative transcends the confines of a conventional racism tale, instead tracing the unique conflux of an individual's life elements. Within these pages unfolds the remarkable journey of an impoverished Indian, granted an opportunity in the land of promise.

The protagonist, Iyengar, finds himself in the Deep South during the early 1970s, an unexpected detour from the coastal elite schools he could have attended. His inaugural job at a historically black college becomes a cornerstone of this chronicle. Here, in the burgeoning field of computer science, he pioneers a program that leaves an indelible mark on Mississippi's future—a formidable accomplishment.

Set against the backdrop of the transformative 1970s, this decade takes on a persona, becoming a driving force and influencer in this authentic narrative. Iyengar's arrival coincides with the first wave of immigrants seeking higher education under LBJ's late '60s reforms. The integration of immigration into the Deep South, his experience at a historically black college as an Indian, and the advent of the computer science era merge into an unparalleled amalgamation. Through early adversities, triumphs, and setbacks, Iyengar embodies the spirit of transcending temporal and spatial confines, fostering an unwavering belief in his impending success.

His journey inspires, spanning five decades and crossing cultural and racial boundaries, catalyzing students of diverse backgrounds. Rooted in the 1970s, this captivating memoir molds Iyengar's illustrious career, offering insights that resonate in our present-day pursuit of equity amid racial bias and social inequality.

"The Circuit Rider of Mississippi" emerges as a vital narrative, illuminating a path forward amidst today's challenges. It underscores the potential for altering destiny, much like the immigrant depicted herein did years ago—proving that transformative change arises not from being shaped by circumstances but from the determined resolve to shape them.

Prologue

In the heart of the Jim Crow South, Dr. S.S. Iyengar stepped off a bus into a world that was starkly different from the one he had left behind in India. Mississippi's rural landscape was deeply divided, marked by the boundaries of segregation, and Dr. Iyengar's assignment was formidable.

Dr. S.S. Iyengar completed his Ph.D. in Engineering in 1973 from Mississippi State University and, shortly after, was offered a position by Dr. Jessie Lewis, Head of the Computer Science Department and Dr. John Peoples Jr., President at Jackson State College. Beginning on January 11, 1974, he took on the role of Circuit Rider, funded by the National Science Foundation, whose mission was to ensure equal access to quality education across Mississippi's colleges and universities. This position would immerse Dr. Iyengar in the challenges and opportunities of teaching computer science to students from Historically Black Colleges and Universities as well as predominantly white institutions statewide, embodying the NSF's vision of "Separate but Equal" education for all. The journey that followed offers a glimpse into the unique obstacles and profound impacts of his work.

As a National Science Foundation-funded Circuit Rider, he was tasked with delivering the future to educational institutions that had barely glimpsed the digital age. His role was far more than a job; it was a mission defined by resilience, one that would challenge his every notion of belonging while igniting a journey of progress in one of America's most segregated regions.

Dr. Iyengar's journey as a Circuit Rider in early 1970s Mississippi embodies resilience, determination, and innovation. Coming from India, he not only faced the adjustment to a foreign land but was met with the harsh realities of racial divides and cultural barriers entrenched in the Jim Crow South. Yet, he pressed forward, fueled by a vision for a future where technology bridged divides, one that would transform the educational landscape for generations to come.

Dr. Iyengar's mission, backed by the National Science Foundation, required physically traversing Mississippi's towns to integrate technology into colleges and universities, especially minority-serving institutions suffering from a lack of essential resources. In these under-resourced schools, his role was critical, providing not just tools but access to a rapidly advancing global community. By enabling students from rural and underserved communities to engage in this new digital age, he created pathways to opportunities that were often limited by their isolation.

Dr. Iyengar opened pathways to opportunities often constrained by the isolation of the communities he served. The environment he encountered was one of entrenched segregation and inequality. As a brown man with a distinct accent, he faced both subtle and overt discrimination, yet he remained steadfast in his mission. His resilience allowed him to stay focused on a greater purpose: fostering both technological and human connections across institutional and racial divides. Working alongside Dr. Pramanik and under the leadership of Dr. Lewis, he helped establish a regional educational network, empowering schools to transcend geographic and racial boundaries and showing that technology could serve as a transformative force for social progress.

Dr. Iyengar's work required more than the installation of a portable computer terminal manufactured by Texas Instruments in the 1970s which had all good features of simple programming. His commitment to connecting institutions meant bridging the physical and technological distances separating these schools. By establishing communication networks and fostering collaborative learning, he facilitated educational advancement, ensuring that Mississippi's institutions could engage in broader national and global academic discourse. His contributions built a foundation for future generations who would continue to benefit from the technological infrastructure he helped establish.

The NSF's grant investment in the Circuit Rider program underscored the importance of ensuring educational equity, and it was instrumental in advancing Dr. Iyengar's work. The Foundation recognized that true progress depended on making technology accessible to all, regardless of geography or economic status. By supporting Circuit Riders, the NSF was not merely advancing technology; it was advancing social opportunity. Dr. Iyengar's journey, therefore, was about much more than bringing technology to Mississippi; it was about enabling inclusion and promoting pathways to success for communities often left on the margins.

Dr. Iyengar's story goes beyond the technical; it is one of breaking down social barriers. The networks he developed were more than just tools of communication; they were lifelines, helping students and educators access a world of opportunities and achieve their aspirations. He didn't just provide technology; he paved the way for a future where technology could be a force for equality, building paths toward educational access and empowerment.

His story reflects the broader challenges of technological growth in America. As advancements rapidly evolved, the gap between connected and disconnected regions grew. The work of Circuit Riders like Dr. Iyengar ensured that rural and underserved institutions would not be left behind. His contributions also serve as a reminder that even as technology moved forward, so, too, could social progress. In a society defined by segregation, connecting people through technology was a quiet act of resistance, proving that progress, whether technological or social, could transcend the divides of the past.

Dr. Iyengar's journey is one of perseverance in the face of profound personal and societal challenges. As an outsider in a deeply segregated society, he faced prejudice but maintained his belief in the transformative power of technology and education. His role as a Circuit Rider in Mississippi exemplifies the lasting impact that one individual can have, as he opened doors for students, expanded the resources of institutions, and contributed to building a more just and equitable society.

The legacy of Dr. Iyengar's work lives on through the networks he established, the students he inspired, and the institutions that continue to benefit from his vision. His path from India to the Jim Crow South is a powerful testament to the importance of resilience, the value of education, and the transformative role that technology can play in creating a better future for all. Dr. Iyengar is now (2024) a highly esteemed computer scientist at Florida International University, played a pivotal role in establishing a groundbreaking Center of Excellence for Digital Forensics, supported by the US Army Research Office. This flagship program, the first of its kind nationwide, is dedicated to advancing the cybersecurity and forensic skills of students from HBCUs and Minority Serving Institutions. By focusing on the development of next-generation technology, the center aims to empower minority students and women, preparing them for impactful careers in cybersecurity and digital forensics.

Acknowledgments

This book would not have been possible without the invaluable guidance, inspiration, and support of many dedicated individuals. It has been a distinct honor and pleasure to work alongside Professor S.S. Iyengar, whose remarkable five-decade journey has spanned continents and touched countless lives. His wisdom, resilience, and dedication to advancing knowledge have deeply influenced every chapter of this book, and his experiences continue to inspire those who seek innovation and impact in their work.

I extend my heartfelt gratitude to all those who contributed their insights, ideas, and suggestions, each adding depth and perspective that enriched this work. Your commitment and enthusiasm have been instrumental in bringing this project to fruition. Thank you for sharing your time, knowledge, and encouragement.

To the many colleagues, students, friends, and supporters who believed in this project and shared in its vision, your encouragement and input have been invaluable. This book stands as a testament to the collaborative spirit that drives forward the pursuit of knowledge and learning. Thank you all for your support, insights, and contributions.

This work was partially supported by the Army Research Office and the NSF funding, and was accomplished under Grant Number W911NF-21-1-0264 and 2018611. The views and conclusions contained in this document are those of the authors and should not be interpreted as representing the official policies, either expressed or implied, of the Army Research Office or the U.S. Government.

Col. Jerry Miller
Yashas Hariprasad
Ranvir Iyengar

Table of Contents

Chapter 1:
Laying the Foundations to a Journey of chasing and realizing dreams

"The future belongs to those who believe in the beauty of their dreams."

— Eleanor Roosevelt

1.0 Introduction

Throughout its extensive history of two hundred and forty-three years, the United States has traversed a remarkable transformation narrative, marked by pivotal moments that catalyzed the emergence of a more inclusive, just, and resilient society. From the brink of dissolution during the wrenching struggle against racial prejudice and the deep-seated ideological chasm between the North and South, epitomized by the crucible of the Civil War in 1861, to the hard-won victory of the Women's suffrage movement, securing voting rights for women in 1919, and the nonviolent resistance that shattered the oppressive barriers of Jim Crow in the 1950s and 1960s, these junctures have symbolized the conflict between progress-seekers and those vested in maintaining a static status quo. The annals of history now acknowledge the year 2020 as yet another transformative juncture not only for the United States but for the global narrative as well. The unprecedented convergence of racial and societal turbulence, amplified by a once-in-a-century global pandemic, cast a profound impact across humanity, transcending backgrounds and circumstances.

Image: American Civil War 1861

Image: Jim Crow 1960

Amid this maelstrom, the citizens of the United States were summoned to exercise their paramount civic duty—casting their ballots to determine leaders at local, state, and national levels, culminating in the election of the President. Hence, the nation found itself at another crossroads, an era necessitating reflection, discourse, deliberation, and potential metamorphosis.

Among the myriad voices that resonated through mail-in and in-person ballots, one belonged to S S Iyengar – a septuagenarian whose personal inflection point had unfolded five decades prior, in September 1970. It was during this pivotal month that he embarked on a journey that would redefine his life, shifting him from the familiar embrace of Bangalore, India, to the sultry ambiance of Starkville, Mississippi.

At the precipice of commencing this deeply intimate odyssey, Iyengar stood encircled by his belongings – his bags meticulously packed – poised to embark on the journey of a lifetime. The doorway of his house framed not just his departure from a physical abode but also his departure from his cherished homeland. Fate, however, interceded in the form of a "well-to-do" friend, whose generosity extended to sharing a taxi ride to the Bangalore airport, an act that proved serendipitous, for without this intervention, the young Iyengar might have missed his pivotal opportunity altogether.

Image: Indian Institute of Science, Bangalore

Having recently attained his Master's Degree in Engineering from the prestigious Indian Institute of Science (IISc) in Bangalore, Iyengar's academic journey had been influenced by luminous mentors such as Satish Dhawan – an architect of India's nascent space exploration program, Sir Chandrasekhara Venkata "C.V." Raman – a pioneering Indian physicist heralded for his discovery of the eponymous Raman scattering effect, and Chintamani Nagesa Ramachandra "C.N.R." Rao – an eminent chemist acclaimed as a recipient of numerous awards, albeit the Nobel prize eluded him. These academic luminaries ignited a spark within Iyengar, compelling him to traverse further realms of education rather than immediately joining the professional workforce. This decision, laden with complexities due to perennial financial constraints experienced by him and his family, was a profound testament to his resolute commitment to his aspirations.

S. S. Iyengar

Prof. Satish Dhawan

Prof. CNR Rao

Sir C. V. Raman

Vice Chancellor Dr. J.M. Vyas

19

Prof. Homi J. Bhabha

The upcoming pages embark on an enthralling narrative that transcends borders and cultures – a story entitled "The Circuit Rider of Mississippi: An Indian Immigrant's Story in the Post-Jim Crow South." This tale is a captivating exploration of one individual's transformative journey, a symbolic and representative passage of the broader societal transformation unfolding in America during this time. As we delve into the various chapters of this narrative, we shed light on the multi-faceted journey of S S Iyengar. This journey was powered by dreams, determination, and an unwavering pursuit of the unknown and limitless possibilities of life. It is a narrative that speaks to the challenges of navigating through uncharted territories – both geographically and metaphorically – and encapsulates the essential qualities of human endeavor, resilience, and the indomitable spirit required to shape one's destiny amidst a world in a constant state of flux.

1.1 Early Life: Seeds in the Soil of Hemmige

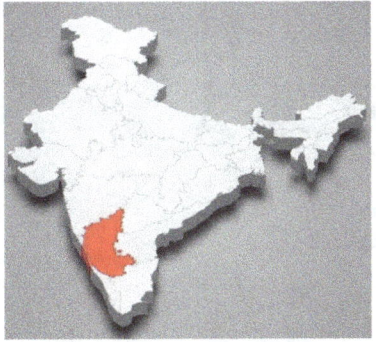

The genesis of our story unfolds in the heart of a rustic Indian enclave known as Hemmige, nestled within the verdant embrace of the Southern Indian state of Karnataka. This modest hamlet stood as a testament to the bygone era, untouched by the luxuries of electricity, automobiles, or even the convenience of water flowing through taps. Born into this humble tapestry, on the cusp of India's newly won independence from British rule, emerged Sundararaj Sitharama Iyengar – the fourth progeny of the union between S.N. Sundararaj Iyengar and Chokkamma.

In its infancy, Hemmige epitomized the quintessential rural Indian landscape of the era, where the rhythms of life were attuned to the simplicity of existence. Here, the absence of electric currents, the lack of running water, and the quietude void of automobile engines were the norms. Through these unpaved streets, the primary mode of conveyance was the age-old act of walking, punctuated occasionally by the rumble of bullock carts on their journey to the bustling markets. Sitharama's entry into this world followed the customary path of his time, transpiring within the confines of a modest mud abode with the assistance of untrained hands and the wisdom of Chokkamma, his venerable great-grandmother.

Image: Rural India

In the backdrop of this quaint setting, Sitharama's father, a luminary in his own right, a chemist by education, navigated life as a divisional assistant within the Mysore State Electricity Board situated in Shimsha. While intellectually engaging, this occupation was financially modest, presenting a perpetual struggle to make ends meet. However, the tides of destiny turned favorably for the Iyengar family in 1949 as they transitioned to the burgeoning metropolis of Bangalore – a city teeming with possibilities and ripe for the enrichment of young minds.

Emerging as the capital of Karnataka State, Bangalore was a vibrant tapestry woven with dreams and potential. The city's pulse was palpable, notably illuminated by its pioneering embrace of electrified streetlights, illuminating pathways that resonated with hope.

Image: Vidhana Soudha, Bangalore, India

Crucially, within this bustling haven rested the esteemed citadel of intellectualism – the Indian Institute of Science, established in 1909, and the Raman Research Institute, forged in 1947 under the stewardship of India's premier Nobel laureate, Sir Chandrasekhara Venkata Raman.

Image: Bangalore, India

As the world around him shifted and transformed, a young S S Iyengar found his life taking shape, influenced by the currents of scientific inquiry and the promise of mathematical discovery. Even in his early years, his prodigious talents in the realms of science and mathematics were evident, and he quickly found recognition within the academic circles of his city.

However, the foundation of Iyengar's convictions lay deep within the teachings of his father and grandfather, wise guardians who instilled in him the values of perseverance and faith. Their belief in the symbiotic relationship between diligence and opportunity, akin to the nonviolent spirit of Mohandas Karamchand Gandhi, that had propelled India towards independence, served as a guiding light for Iyengar in his personal aspirations.

These principles gained even greater significance as India's civil rights crusaders began to make their voices heard, demonstrating the power of resolve against the formidable might of the British Empire. For Iyengar, the echoes of their triumphs resonated deeply, inspiring him to pursue his own journey of self-discovery and independence within uncharted domains.

1.1.1 Early Education: The Illuminating Path to Knowledge

The tapestry of young Iyengar's life was woven with threads of intellect, perseverance, and insatiable curiosity right from the inception of his educational journey. His initial foray into the realm of learning commenced within the walls of Nagappa Block Government Primary School in 1953. Here, amid the hum of eager young minds, his brilliance shone bright, prompting an exceptional offer – the opportunity to leapfrog from the first to the second grade, a testament to his innate acumen.

Diligence and inquiry were his constant companions, and the question "Why?" became his hallmark inquiry. This quest for understanding spurred him to explore diverse avenues, igniting his intellectual wanderlust even as he transitioned to middle school. The constraints of limited resources and food did little to dampen his ardor for excellence. Instead, they seemed to fortify his resolve, propelling him to ascend the academic ladder with a predilection for mathematics – a discipline that not only enabled him to unravel complexities but also endowed him with the art of analytical thinking and astute reasoning.

In this journey of knowledge, Iyengar emerged not only as a solitary scholar but as a benevolent guide for his peers. His generosity in assisting fellow students, a quality he cherishes to this day, reflected his understanding that collective growth fosters an environment of shared brilliance. Yet, amidst these laurels, a persistent inner voice spurred him with the words

"You can do better!" — a voice that refused to let him rest on his laurels, an ever-present impetus for further achievement.

Image: **State Central Library, Bangalore, India**

Despite limited financial resources, Iyengar did not allow himself to be held back. Instead, he channeled his ingenuity and resourcefulness to seek out knowledge through frequent visits to local libraries. Despite not possessing any personal copies of books, he immersed himself in a diverse range of literature spanning various disciplines. His insatiable hunger for knowledge was never satiated, and each book he read further enriched his understanding of society and ignited new passions, contributing to his holistic comprehension of the world.

As he progressed to higher education, Iyengar's pursuit of excellence found a nurturing environment at Seshadripuram High School. Here, he was surrounded by a select cohort of peers and a faculty distinguished by superior teaching and unwavering discipline. Within this conducive atmosphere, Iyengar thrived and delved deeper into the realms of Science and Mathematics. His passion for these fields only grew, and he developed an ever-more profound appreciation for their intricacies and complexities.

A pivotal moment arrived at the tender age of fifteen when Iyengar triumphantly completed the requirements of his high school education, earning the accolade of a top-ranking student across various subjects. This achievement not only marked a personal milestone but also laid the foundation for the extraordinary journey that lay ahead.

The halls of learning had become the arena for young Iyengar's growth – an environment that nourished his intellect, honed his analytical prowess, and kindled a yearning for knowledge that would serve as the driving force propelling him toward uncharted horizons.

1.1.2 Navigating Crossroads: Balancing Family Responsibilities and Educational Aspirations

With the chapter of high school neatly folded, Iyengar confronted a momentous juncture that could potentially reshape the course of his life. The flames of ambition burned fervently within him, urging him to pursue further education. Yet, the backdrop of his familial circumstances cast a shadow of uncertainty. The weight of responsibilities at home stood in stark contrast to the soaring trajectory of his aspirations, creating a delicate balancing act that demanded ingenuity, perseverance, and a dash of audacity.

The yearning for continued learning was undeniably potent, but the stark reality was that his family's financial circumstances loomed as a formidable barrier. The prospect of funding his tuition seemed remote, as his own pockets were devoid of resources. The choice of taking up immediate employment was a tempting avenue, albeit one that would inevitably shutter the doors to his dreams.

The path to adulthood was paved with complexities that cast the weight of reality upon his youthful shoulders.

Undeterred by the challenges that lay before him, the resolute young Sitharama devised a strategy that hinged on a bold gamble—one that would involve not only seeking financial support from his mentors but also swaying his father's heart and mind. Armed with his determination and an eloquent plea, he embarked on a journey to present his quandary to his teachers. This endeavor required not just courage but an unwavering belief in the power of education to transform his destiny. His appeal to his educators was more than a mere request for monetary assistance; it was an impassioned entreaty for the realization of his aspirations. With conviction, he communicated his deep-seated need to bridge the gap between his yearning for knowledge and the limitations posed by his circumstances. His plea carried within it the hopes of a young mind eager to soar beyond the confines of his current situation to carve a path illuminated by the light of learning.

This audacious maneuver yielded remarkable results – his mentors responded with a blend of empathy and support, recognizing the sincerity that colored his quest for knowledge. Armed with their endorsement, he embarked on a dual mission. The first was to secure his father's approval, which required finesse in articulating the intrinsic value of education and aligning his aspirations with his father's hopes for a brighter future. The second was to pursue a seat at the prestigious National College at Basavangudi, a beacon of learning nestled within the heart of Bangalore.

Image: National College, Basavangudi, Bangalore, India

Sitharama's determination and his teachers' endorsement prevailed, and he found himself within the hallowed halls of National College. This phase marked not just the continuation of his education but the embodiment of his resilience and ability to navigate complex intersections with unwavering resolve. The pivotal decision to seek both financial aid and familial approval served as a testament to his foresight and his unyielding pursuit of enlightenment.

As he stepped onto the campus of National College, he carried with him not only his textbooks but also the culmination of his aspirations and his triumph over adversity. The journey ahead was one laden with challenges, yet it was also suffused with the radiance of a young mind refusing to be tethered by circumstances. The story of S S Iyengar was evolving, with each chapter etching the contours of a life fueled by dreams and the courage to chase them against the odds.

1.2 College Life: The Dawn of a New Epoch for Iyengar

Stepping through the gates of National College in Basavangudi marked not merely the beginning of an academic journey but the dawn of an intellectually invigorating era for S.S. Iyengar. Surrounded by an air of academic brilliance and an environment teeming with scholastic pursuit, Iyengar found himself in the company of individuals who were more than just fellow students—they were beacons of inspiration and role models to many aspiring minds. Among these luminaries, one figure stood tall and left an indelible mark on Iyengar's formative years—Dr. H. Narasimhiah, the revered Principal of National College. Armed with a doctorate in nuclear physics from Ohio State University, Dr. Narasimhiah was more than just a scholar; he was a visionary leader. His presence alone infused the corridors of the college with a palpable sense of purpose and ambition, encouraging students to not only dream big but also pursue their academic goals with unyielding determination.

Under Dr. Narasimhiah's leadership, National College became a fertile ground for intellectual nourishment. The college's vast library, with its boundless collection of books and resources, became Iyengar's sanctuary, a space where his educational aspirations could grow unhindered. The treasure trove of knowledge housed within those walls, combined with the intellectual vibrancy that permeated the campus, made his daily 12-kilometer trek from home to college an enjoyable pilgrimage. Each step toward National College was a step toward enlightenment, and Iyengar embraced the journey with a thirst for knowledge that knew no bounds.

But it wasn't just the academic infrastructure or the environment that fueled his curiosity. National College regularly opened its doors to some of the greatest minds in the world, organizing guest lectures that provided students with exposure to the latest developments in various fields. It was during one such lecture that Iyengar's life was profoundly altered. When Nobel laureate C.V. Raman took the stage, his words cast a transformative spell over Iyengar. Raman's discourse on science and research ignited a passion, kindling a newfound desire to explore the frontiers of engineering and innovation.

However, the road to becoming a researcher in India at that time was fraught with challenges. Limited resources and opportunities made it difficult for budding scientists and engineers to pursue cutting-edge research within the country. Yet, Iyengar saw these challenges not as roadblocks but as motivators, driving him to find a path to contribute meaningfully to his nation's development. Inspired by Raman's words and driven by his ambition to make a difference, Iyengar turned his focus toward engineering as a means of bringing about progress in his homeland.

From that pivotal moment, Iyengar's course was set. With the odds stacked against him, he sought out scholarships, took on educational loans, and worked tirelessly to fund his education. His efforts paid off, and he eventually secured a place at the University Visvesvaraya College of Engineering, a significant step in his journey toward becoming one of the leading figures in the world of computer science and engineering.

The lessons learned at National College—inside and outside the classroom—became the bedrock of Iyengar's academic and professional life. It was there that he not only honed his intellect but also developed a resilience that would carry him through the challenges of his future career. His time at National College, under the mentorship of Dr. Narasimhiah and influenced by the brilliance of minds like C.V. Raman, laid the foundation for a life devoted to research, engineering, and the pursuit of knowledge.

In retrospect, National College was not just an educational institution for Iyengar; it was the crucible in which his intellectual and personal philosophies were forged. The experiences, relationships, and opportunities he encountered during his time there set the stage for a remarkable career, one that would ultimately shape the future of computer science and engineering on a global scale.

Iyengar recalled an incident when he was young and hungry on Jan. 12, 1960. He had gone to IISc to listen to a lecture by Sir. C. V. Raman in the Raman Institute of Science. Raman was an excellent speaker full of confidence and an aggressive personality.

After the seminar, he was hungry and went to a marriage ceremony (uninvited) in Malleswaram, near the 8th cross. He was sitting with others to have his dinner.

At mealtime, a person serving the food accidentally asked Sitharama to which side (groom/bridegroom) I belonged.

"I was not able to respond properly, and after a few minutes, I left the place without finishing my food."
-S.S. Iyengar

With bated breath, a nervous Dr. Iyengar waited for a thunderous reaction from the audience. However, he was in a big surprise.

Iyengar was an ardent believer in family values and family affection. He loved his parents and believed in the adage that one father is more than a hundred schoolmasters. But during his tender age, his parents and relatives did not touch the right chord. Hope often rises out of the ashes of despair.

"The more energy i spend in elevating and serving others the more energy flow into me".
-S.S. Iyengar

Image: An Enduring Bond - S. S. Iyengar Breakfast with Father, 1995

In 1963, a pivotal chapter in the life of S.S. Iyengar began as he embarked on a journey that would define his professional trajectory for decades to come. Armed with an unquenchable thirst for knowledge and an unwavering determination to succeed, Iyengar set his sights on earning a bachelor's degree in mechanical engineering. This was no easy path, and the years that followed were marked by relentless hard work, long hours of study, and a deep commitment to excellence. By 1968, after five years of tireless dedication, Iyengar had met all the requirements for his degree. His professors, who had closely observed his journey, were unanimous in their praise, recognizing him not only as a brilliant student but as someone who consistently stood head and shoulders above his peers. Iyengar's academic achievements were nothing short of exceptional, as he secured first-class distinctions in all five engineering examinations, a feat that placed him at the top of his graduating class.

Yet for Iyengar, these accolades were not an end in themselves. The honor and recognition, while deeply appreciated, were not the culmination of his efforts but rather stepping stones toward even greater aspirations. Iyengar was never one to rest on his laurels. His modus operandi was defined by a restless desire for continuous improvement and a vision that always stretched beyond the immediate horizon. As soon as he earned his bachelor's degree, he set his sights on a new goal: a master's degree in engineering. By June 1968, mere months after completing his undergraduate studies, Iyengar had already stepped into the prestigious Indian Institute of Science (IISc) in Bangalore, a renowned institution that stood as a beacon of advanced scientific and engineering education in India.

At IISc, Iyengar found himself immersed in an environment that was intellectually rigorous but inspiring. Among the many brilliant minds that shaped his time there, one figure stood out as a profound influence on his academic journey—Prof. Sathish Dhawan. Fresh from the halls of the California Institute of Technology, Dhawan was a luminary whose expertise in aerospace engineering and visionary leadership later helped shape India's space program. For Iyengar, Dhawan's lectures were more than just academic sessions; they were moments of enlightenment that expanded his understanding of engineering and its potential to drive societal progress. Dhawan's approach to teaching offered a reprieve from the rigors of constant study, but more importantly, it fanned the flames of Iyengar's intellectual curiosity and stoked his desire to push the boundaries of what he could achieve.

As Iyengar delved deeper into his studies, his vision for the future began to crystallize. It was during this time that he realized the importance of educational independence and the need to explore opportunities beyond the borders of India. Although India had produced some of the brightest minds in the world, Iyengar believed that in order to fully unlock his potential and contribute to his nation's progress, he needed to seek out new perspectives, advanced research facilities, and cutting-edge technologies that were only available abroad. The United States, with its renowned universities and research institutions, became his target. He saw this as the key to not only his personal advancement but also as a way to contribute to the larger goal of propelling India toward prosperity and global recognition.

For Iyengar, education was never merely about personal achievement; it was always tied to a larger purpose. He held a deep-seated belief that the knowledge and skills he would gain from studying abroad could be the catalyst for meaningful change in his homeland. India, during this period, was grappling with issues of poverty, underdevelopment, and a lack of access to modern technology. To Iyengar, education was the answer—a means to uplift his fellow countrymen, instill hope, and contribute to building a future where India could stand tall among the world's leading nations. He envisioned a future where the expertise he gained abroad could be used to address the challenges facing India, from economic disparity to technological stagnation.

This conviction was not a fleeting thought but a guiding principle that motivated Iyengar through the long hours of study, the sacrifices, and the challenges he faced along the way. He was driven by an unshakable belief that his success was intertwined with the success of his homeland. The idea that education could be a force for national development had been nurtured by the mentors and visionaries who had guided him thus far—figures like Prof. Dhawan, who had themselves experienced the transformative power of education and innovation. They instilled in Iyengar a sense of responsibility to use his talents for the greater good, a responsibility that he embraced wholeheartedly.

By the time he completed his master's degree, Iyengar had already begun laying the groundwork for the next phase of his journey—a phase that would take him beyond India's borders to the United States. Armed with his academic credentials, unwavering determination, and a vision for the future, he prepared to embark on this new chapter, knowing that the challenges ahead would be great, but so too would be the opportunities. The trajectory of S.S. Iyengar's life and career was propelled by a confluence of factors: his personal drive, the support and guidance of brilliant mentors, and a vision for how education and innovation could be harnessed to transform not just his life but the future of his homeland.

As Iyengar set his sights on the U.S., he was not merely chasing personal ambition; he was stepping into a larger role as a torchbearer for a generation of Indians who believed in the power of education to drive societal change. This journey, marked by perseverance, intellectual rigor, and a profound sense of purpose, would ultimately shape his legacy as a pioneer in the fields of computer science and engineering and as a visionary leader dedicated to the advancement of knowledge and the betterment of his nation.

1.3 Higher Education - Indian Institute of Science (IISc): A Luminous Chapter (1968 – 1970)

The chronicles of Iyengar's higher education unfolded within the hallowed corridors of the Indian Institute of Science (IISc), an institution renowned as one of India's premier bastions of learning. Amidst its august premises, a new phase of intellectual growth blossomed under the tutelage of luminaries who left an indelible mark on his journey.

Image: Indian Institute of Science (IISc), Bangalore, India - An institution renowned as one of India's premier bastions of learning

During his tenure at IISc, one figure emerged as a guiding beacon in Iyengar's trajectory - Dr. Satish Dhawan, a luminary who not only inspired Iyengar but became his mentor. Having recently returned from the illustrious California Institute of Technology, Prof. Sathish Dhawan infused Iyengar's academic journey with a blend of inspiration and respite. Through Dhawan's visionary lectures, Iyengar found moments of solace, temporarily liberated from the rigors of ceaseless study. Within this nurturing environment, Iyengar was not just cultivating knowledge; he was cultivating dreams – visions of a rejuvenated India intertwined with his own prospects within that vision. Reflecting on this period, Iyengar fondly reminisces, "Sathish Dhawan, Director of the Indian Institute of Science when I was a student, was my respected mentor. It was he who advised me that I needed to do the Ph.D. and suggested that the United States might be the best place for me."

A legacy of this mentorship is the NIAS-MAIYA Prodigy program, a testament to Iyengar's commitment to nurturing future generations. In collaboration with the National Institute of Advanced Sciences (NIAS) in India, this program grants promising high school and college students financial scholarships and mentoring, empowering them to embrace their aspirations. The seeds of Iyengar's passion for education were nurtured within these very corridors, and the harvest of his dedication continues to inspire generations.

However, the passage of time kindled a restlessness within Iyengar. Two years within the sanctums of IISc had cultivated not only knowledge but also an eagerness to explore new horizons. His gaze turned westward, and he embarked on a journey to chase his Ph.D. dream in engineering. Despite his aspirations, the constraints of funding loomed heavily. Among his fervent applications, it was the embrace of Mississippi State University that provided a lifeline – a full scholarship that would fuel his quest for knowledge.

As Iyengar embarked on a new path, leaving behind the familiar comforts of his homeland, he was at the cusp of a transformative journey. Meanwhile, in a geographical location far from his own, the state of Mississippi was grappling with its own historical struggles. The echoes of the American Civil War still resonated, and Mississippi was burdened by the weight of its past. Back in the 1800s, Mississippi had been at the forefront of cotton production, a thriving industry fueled by sprawling plantations and the labor of enslaved black workers. However, the aftermath of the Civil War in 1865 left Mississippi in shambles, not just culturally and intellectually but also economically.

As Iyengar pursued his educational ambitions, he crossed paths with Mississippi at a critical juncture in its own journey. The state was still weighed down by the remnants of its past, struggling to move forward. The divergent paths of Iyengar and Mississippi, one filled with hope and promise and the other with the weight of history, created a powerful backdrop against which the transformative journeys of individuals and societies unfolded. It was this convergence of two seemingly disparate narratives that ultimately created a unique tapestry, interweaving destinies in the most unexpected ways.

Lessons Learned

"Education is the manifestation of perfection already in men."

-Swami Vivekananda

India has long been known as a land of knowledge and home to some of civilization's greatest teachers. One of these great teachers was Swami Vivekananda who had significant influence in America in the late 1800s. He traveled widely and lectured throughout the United States. Swami Vivekananda's desire was to instill a sense of confidence in every person that they were already blessed with infinite potential.

For young Sundaraja Sitharama Iyengar, there was no doubt in his mind that he was blessed with infinite potential. Each level of his educational journey gave him greater confidence, reinforcing his blessing.

Iyengar recognized early in his academic career that continuing to open the blessings of infinite potential came through hard work and dedication. Each step of his educational process unlocked yet another element of his genius. As Thomas Edison stated in a letter in 1927, "Genius is one percent inspiration and ninety-nine percent

---

perspiration." Iyengar would adopt not only this motto, but this same work ethic throughout his career.

Another of S.S. Iyengar's lessons learned can be summed up in a quote by Leonardo da Vinci.

"It had long come to my attention that people of accomplishment rarely sat back and let things happen to them. They went out and happened to things."

-Leonardo da Vinci

And so it was that S.S. Iyengar set his sights on continuing his education in America, a pathway that would set his destiny.

Iyengar wanted to experience the depth and breadth of America and so chose Mississippi State University (MSU) in America's heartland, as his university of choice. Originally established as a land-grant college in 1878, *The Agricultural and Mechanical College of the State of Mississippi,* redefined their education and research programs in the 1950s, adding doctoral degree programs in 1951, and the College of Arts and Sciences in 1956 (Mississippi State University, 2024). Mississippi and all the United States were improving the American lifestyle through science and education.

Below the surface, however, the cauldron of civil rights boiled like a volcano ready to explode. Black Americans were rejecting the "Jim Crow laws" springing from the doctrine of "separate but equal" which had been in effect since 1896. A 1954 U.S. Supreme Court decision in *Brown v. Board of Education* effectively ended legal racial segregation in public schools in the U.S., although many schools remained segregated.

In 1955 in Montgomery Alabama, Rosa Parks refused to give up her seat to a white man and sit in the colored section in the back of the bus. During the ensuing decade little progress was made in desegregation in the "deep South" of the United States, despite the nonviolent efforts of Dr. Martin Luther King, Jr. and other black pastors and civil rights leaders. By 1961, "freedom riders," black and white activists took up the mantle against segregation by taking bus trips throughout the South, sitting wherever they pleased, attempting to eat at segregated lunch counters, and using the "whites-only" restrooms. By 1963 America was on the verge of extreme civil rights violence. Government officials raced to provide a solution.

In 1964 the United States enacted a landmark civil rights and labor law outlawing discrimination or segregation based on race, color, religion, sex, or national origin. Shortly thereafter, in 1965 the first black student, Richard Holmes, quietly enrolled and began classes at Mississippi State University without fanfare, despite violent protests that continued to occur in the U.S. in 1965. Race riots finally erupted in more than 110 U.S. cities on April 4, 1968, following the assassination of civil rights leader Dr. Martin Luther King Jr. The riots continued throughout the summer, eventually subsiding while the unresolved transformation of race relations in the U.S. continues slightly below the surface to the present day.

Iyengar was about to enter a unique period of transformation in America which would forever change the way he looked at the world and the role of education. He would not, "sit back and let things happen…" but would actively intercede using the new technology of computer science to effect change.

Chapter 2:
Journey to the Promise Land

"Courage is not the absence of fear, but the triumph over it. The brave man is not he who does not feel afraid, but he who conquers that fear."

— Nelson Mandela

2.0 Journey Towards Mississippi – 1970

For Iyengar, the bedrock of his journey was rooted in the profound value he placed on family bonds and the embrace of cherished affections. A deep reverence for his parents was etched within him, informed by the belief that a single father's guidance holds the weight of a hundred schoolmasters. However, the tapestry of his relationship with his family and relatives bore the marks of discord during his formative years. Yet, within these turbulent currents, he found the seeds of hope, as adversity often sows the soil for growth.

Iyengar's belief system revolved around the potent principle that by dedicating himself to uplifting and serving others, he could rejuvenate his own spirit and energy. His expedition to the United States was no mere linear journey; it was a narrative of twists and turns that had its inception long before he even stepped onto American soil to chase his aspirations. Amidst these complexities, the journey wasn't a solitary pursuit; it was a symphony orchestrated by a myriad of interconnected experiences, each composing a chapter that paved the way for his transformative odyssey.

The journey to embark on the path of pursuing higher education in the United States was a saga that began with the interplay of familial dynamics. Through adept persuasion, Iyengar engaged in a delicate dance of dialogue, ensuring that his family's concerns were addressed and their blessings secured for his ambitious venture. This intricate process served as the initial chiseling of his path, fostering the alignment of his personal aspirations with the collective hopes of his loved ones.

As the stage was set for his international expedition, Iyengar encountered several obstacles, each presenting an opportunity for growth. The labyrinthine process of obtaining a passport, a tangible testament to his embarkation on a journey of self-discovery, was a precursor to the labyrinth of experiences that lay ahead. The visa application process, replete with its complexities, became a rite of passage that tested his patience and determination, affirming his commitment to his dreams.

In the midst of these bureaucratic navigations, the simple act of capturing his first official photograph – a snapshot that would come to symbolize his identity on foreign shores – carried a significance beyond its physical form. It embodied his metamorphosis, a visual encapsulation of the dreams and aspirations that fueled his voyage.

The myriad challenges that punctuated his path served as a formative prelude to the tempestuous times that awaited him upon his arrival in the United States. Each instance was a lesson in resilience, resourcefulness, and adaptation, imparting invaluable skills that would serve as armor in the face of the unfamiliar.

Thus, the narrative of Iyengar's journey toward Mississippi was not a linear progression; it was a tapestry woven with the threads of determination, familial bonds, and the tenacity to overcome obstacles. In each of these instances, he planted seeds that would eventually bear fruit, shaping the course of his narrative and embodying the ethos guiding his footsteps on American soil.

2.1 Navigating the Passport Maze: Divine Intercession

The canvas of Iyengar's journey to the United States unfurled upon the intricate tapestry of obtaining a passport – a novel concept that heralded both possibility and challenge. Within the fold of this passport lay the key to his imminent sojourn, a symbolic gateway and a tangible emblem of his identity in a distant land. Yet, this endeavor was not devoid of its tribulations, and the very first stroke in this chapter of his expedition was painted by a photograph.

Figure: Obtaining Indian Passport

The photographic encounter marked a pivotal moment of Iyengar's life—his very first brush with this visual medium, a captured memory that would forever be enshrined in his heart. On June 15, 1970, at a juncture bathed in anticipation, armed with the requisite photograph and supporting documents, Iyengar stood poised at the Bangalore office, ready to submit his application. However, the realm of bureaucracy had its own script to unveil.

As fate would have it, Iyengar's journey toward his goal was unexpectedly impeded by an additional demand. This new requirement mandated a financial certification that must be duly endorsed by an officer holding the prestigious IAS cadre position. This condition, which seemed to appear out of nowhere, was akin to a curveball in a well-choreographed dance. It left Iyengar in a state of disillusionment, feeling lost and uncertain of how to proceed.

Despite his father's affiliation with the office, Iyengar found himself unable to secure the influential signature required for the certificate. His familial network lacked the necessary connections, and even his relatives remained conspicuously absent, leaving him adrift amidst a sea of challenges. The situation seemed bleak, and Iyengar felt like he was fighting an uphill battle.

But in the moments of despair, Iyengar's faith remained unwavering. He turned to prayer, seeking divine intervention and a guiding hand to part the clouds of uncertainty. And as if scripted by destiny itself, the tides turned in his favor. It was in the quietude of a Bangalore evening, against the backdrop of the IISc library, that Iyengar's life would intersect with a fateful stranger. This encounter would change the course of his journey and lead him one step closer to his ultimate goal.

Mr. Reddy, an Estate officer at IISc, emerged as a serendipitous figure in this unfolding tale. He traversed the campus on his scooter, and in the enigmatic dance of fate, he chanced upon Iyengar in his contemplative solitude. Evidently moved by Iyengar's demeanor, Mr. Reddy halted his scooter, extending a bridge of conversation. This encounter served as a crossroads where the trajectories of two lives converged.

Empathy was the currency of this chance meeting as Iyengar unfolded his predicament to the curious officer. Mr. Reddy, without hesitation, extended his benevolent hand, promising his aid and suggesting a rendezvous for the following day. And true to his word, the sun dawned on a new day that bore testament to the power of human compassion. With Iyengar perched on the back of Mr. Reddy's Vespa Scooter, they journeyed to the Vidhana Soudha's office, where an auspicious signature from a distinguished IAS officer was procured.

This harmonious interplay of chance and compassion emerged as an unforeseen miracle, a divine intercession that not only circumvented bureaucratic hurdles but also etched an indelible memory within Iyengar's heart. In the tapestry of his journey, this fleeting meeting with Mr. Reddy stood as a testament to the benevolence that often illuminates the most challenging paths. Yet, fate is not always kind enough to weave threads into lasting bonds. Despite his many subsequent visits to the sanctuary of IISc, Iyengar's path never crossed with Mr. Reddy's again, leaving a sense of longing to express his profound gratitude for the beacon of light that had guided his way.

2.2 Navigating the Visa Voyage: The Dawn of Opportunity

In the early days of June 1970, a pivotal chapter in Iyengar's journey unfurled as he embarked on a pilgrimage to Madras (now Chennai) – a voyage undertaken with a singular purpose: the attainment of a coveted US Visa from the American Consulate office. This juncture was emblematic of the culmination of his efforts, a testament to the scholarship he had secured during his tenure at IISc, an investment that would catalyze his flight towards new horizons.

Figure: Obtaining a USA Visa

Endowed with the promise of scholarship yet tethered by financial constraints, Iyengar's expedition bore the spirit of resourcefulness. Armed with the scholarship funds, which encompassed Visa expenses and travel fares, he embraced an overnight train journey to Madras. The rhythmic cadence of the train's movement mirrored his anticipation, each passing moment a heartbeat closer to his encounter with destiny.

As the sun painted the sky with hues of dawn, Iyengar found himself on Madras soil, the city unfolding its welcoming arms to this aspirant traveler. With the consulate appointment scheduled for 10 AM, he strode towards the consulate, a space laden with the weight of aspirations, dreams, and the echoes of countless narratives seeking validation.

The consulate was shrouded in anticipation as Iyengar joined the queue, his patience bolstered by the promise of the opportunity that lay ahead. The hours stretched long, the minutes slow-danced to the rhythm of the sun's ascent, and the momentous 10 AM approached with tantalizing gradualness.

Yet, the initiation of the visa interview was not without its bumps. As Iyengar faced the Visa Officer, the symphony of English accents met his ears, a harmonious fusion of diverse linguistic tones. This amalgam of sounds, however, posed a challenge to Iyengar's understanding. Accents unfamiliar to his ears intermingled, forming a linguistic mosaic that he struggled to decipher.

In the face of linguistic hurdles, Iyengar's response was a symphony of nods – a nonverbal affirmation that danced to the tune of universal understanding. The knots in his linguistic journey, though perplexing, did not eclipse the brilliance of his academic trajectory. His sterling records and the beacon of the scholarship he had earned projected a narrative transcending language, speaking volumes of his potential and dedication.

This narrative, resplendent with academic merit and scholarly ambition, was the chord that resonated within the chambers of the Visa Officer's deliberation. A 20-minute interaction unfolded, marked by the silent symphony of nodding affirmations and the unspoken exchange of aspirations. As the minutes wove their tapestry, the Visa Officer's judgment crystallized a verdict that transcended the linguistic labyrinth – a US Visa was bestowed upon Iyengar.

With this symbolic seal of approval, Iyengar's expedition to the United States was no longer a distant vision; it was a path illuminated by the guiding light of opportunity. This chapter, defined by queues and nods, encapsulated not only the complexity of bureaucratic processes but also the power of resilience and aspiration. As the consulate's doors closed

behind him, Iyengar walked away with a prized possession –
the visa that would transcend borders and mark the
commencement of his transformative odyssey.

2.3 Nightmares and Lessons: A Dreamer's Odyssey

Amidst the meticulous preparation of applications for
numerous universities in the United States, a recurring
nightmare took root in Iyengar's subconscious, a journey to his
past within the hallowed halls of IISc. This nocturnal odyssey
propelled him back to the Spring of 1970, where he found
himself ensnared in a course taught by Prof. M. V.
Narasimhan. Within this reverie, a haunting narrative unfolded
– a narrative depicting unfavorable scores in assignments and
tests, a perilous trajectory that loomed over his ambition like a
storm cloud. This distressing dream cast a foreboding shadow,
potentially threatening to sever the threads that bound him to
his Ph.D. aspirations in the United States.

The weight of this dream hung heavily upon his psyche, an
unwelcome companion in the pursuit of his dreams. It was not
merely a fleeting phantasm; it was a specter that seemed to
burrow its roots into the fertile soil of his consciousness.
Prayers and petitions, though persistent, held no dominion
over the persistence of this recurring dream. It was in the realm
of slumber that his fears found a voice, a landscape where his
worries materialized with haunting clarity.

One fateful morning, as the clock struck 5 AM, Iyengar was
roused from this unsettling reverie, awakening in the clutches
of his nocturnal phantasm. The aftermath of the nightmare
lingered like a phantom, its grip refusing to loosen. Yet, his
experience in that early morning was not solitary; it was

observed by his grandmother, a silent spectator to the tempest that raged within his mind. For almost 40 minutes, she bore witness to his tremors, a testament to the profound impact this dream held over him.

As the aftermath of this unsettling experience, Iyengar's interpretation took root – a nightmare that transcended its role as mere mental turbulence. It was, for him, a lesson woven into the fabric of his subconscious, a lesson that resonated with the chords of resilience and perseverance. Amidst the ominous imagery, he discerned a message that adversity, often cloaked in the garb of distress, carries within it the seeds of resilience. The very dreams that threaten to dismantle aspirations can serve as a catalyst for fortitude, urging the dreamer to fortify their resolve.

This experience became a lodestar that guided Iyengar's perspective, shaping his perception of challenges as crucibles of growth. He found himself imbued with a deeper understanding – that the crucible of adversity, no matter how daunting, is often a breeding ground for strength. As he reflects upon this formative episode, his words echo with wisdom, "Let us not overlook the fact that amidst catastrophic changes, there lies a silver lining beneficial to the pursuit of our dreams."

Thus, the nightmare that had held him captive in its grip transformed into a teacher, imparting a lesson that would accompany him on his journey across oceans and into uncharted territories. In the corridors of dreams, Iyengar discovered the art of resilience, a treasure hidden within the labyrinth of adversity.

Note: According to Hindu astrological science, dreams and signs manifest on the palm based on past deeds and indicate the good and bad.

(Source https://ijip.in/wp-content/uploads/ArticlesPDF/article_3f1b04e59834e0680497f01adadf84cd.pdf)

2.4 Embarking on the Path of Dreams: A Voyage of Transformation

Iyengar's story reached a crucial turning point - a moment that held the power to shape his aspirations into reality, to take his dreams to the next level and set foot on the soil of promise. However, the journey from India to the United States during the 1970s was no easy feat; it was a tapestry woven with complexities and challenges, a voyage that demanded both determination and resilience.

In a landscape where direct international flights from Bangalore were non-existent, Iyengar's journey unfolded with a mosaic of domestic flights and international connections. At the young age of 22, with dreams in his heart and unwavering resolve, he embarked on a domestic flight from Bangalore to Bombay (which is now Mumbai), poised to take his inaugural international flight from Santacruz Airport. This phase marked the opening act of his transformative journey, a narrative interlaced with the threads of support and determination. Despite the challenges that lay ahead, Iyengar was undeterred and ready to take on whatever came his way.

Central to this chapter was Iyengar's brother, Mr. Govindaraja – a beacon of unwavering support and an embodiment of familial solidarity. Gathering 10,000 INR from various sources, Govindaraja championed Iyengar's aspirations, propelling him towards his American dream. In a time where the exchange rate danced to the tune of 1 USD equating to 4.50 INR, the cost of travel to the US ranged between 4000 INR and 4200 INR. The funds that Govindaraja marshaled stood as a testament to the power of familial bonds, a lifeline that enabled Iyengar to purchase his first pair of shoes, a suit, and essential accouterments.

As the pieces fell into place, Iyengar's heart swelled with pride – a sense of accomplishment that had been forged through resilience, determination, and an indomitable spirit. This chapter was a mosaic of choices – choices to stand firm in the face of setbacks, to forge ahead with a spirit untamed by adversity. It was a juncture that could have easily been a crossroads of defeat, yet Iyengar's unwavering optimism and resolute spirit painted it as a triumphant milestone in his journey.

Dream before I left Bangalore to The Promised Land

"I had a dream that I was taking a few courses at IISC and not doing well in the spring semester of 1970, and One professor, Dr. M. V. Narasimhan, had failed me in one course that he was teaching. This means my going abroad was shattered. My great aspirations and exploring new roads in my life were going out of my hand, and I felt my Lord Lakshmi Narasimha did not help me in this process."

"This was early morning 5 am. I woke from my dream when my grandmother woke me up as she was telling me that I was crying in my dream for more than 40 minutes."

"Let us not forget to remember that these catastrophic changes have something good to offer in the context of pursuing the dreams"

"I get this dream quite often in my life. What has taught me Prowess and dedication that made me preserve through my obstacles, and I have shown that the virtue of adversity is fortitude"
-S.S. Iyengar

.

D-Day dawned – the day of his departure, the day to bridge the distance between aspirations and reality. Amidst the embrace of his parents and relatives, Iyengar set forth from his residence in Sriramapuram, Bangalore, stepping into the realm of the unknown. A taxi ride to the airport was not just a journey in the distance; it was a journey into the realm of new

40

experiences, a testament to the progression he was forging. Within the secure confines of the airport, the first flight on his path unfurled with a delay of five hours – a testament to the intricacies of travel during that era. In the shadows of this delay, Iyengar grappled with his emotions – anticipation intertwined with restlessness, and the specter of homesickness began to rear its head. The solitude of the hotel room in Excelsior Hotel, provided for international passengers of Kuwait Airlines, became both a luxury and a source of unease. In this unfamiliar space, Iyengar confronted the internal struggle of facing the unknown, battling thoughts of retreat and abandonment. The solitude, though cloaked in comfort, amplified the pang of homesickness.

Within this tempest of emotions, questions surfaced – profound queries that echoed within his mind:

- Is the strength of an individual innate or nurtured by circumstances?
- Why does instrumental power tend to favor the affluent, the influential, and the connected?
- Does an enigmatic force, often termed good luck, underlie the orchestration of events?

Amidst this tapestry of contemplation, an announcement sliced through the air – a call to action from the Kuwait Airlines counter, summoning passengers bound for the US. In the wake of this call, Iyengar's moment of truth materialized. He answered the call, boarding the bus that would ferry him to the aircraft – a vessel that would navigate him through Kuwait, London's Heathrow Airport, and finally to the shores of New York's Ellis Island.

The flight that bore him across miles was more than a trajectory; it was a journey within. As the plane soared amidst the clouds, Iyengar grappled with both turbulence and resolve. He was not merely crossing geographical boundaries; he was traversing the depths of his own resilience, fortifying his spirit against the currents of doubt. This chapter was not just the first leg of his journey to a foreign land; it was a prologue to the story of self-discovery, a tale of transformation that would unfold against the canvas of the United States.

> " The lower middle-class family into which that I was born did not deter me from aiming high. A strong zeal to do things easily what others found difficult was the talent I had acquired by dint of hard work"
> **-Iyengar**

2.5 A New Chapter Unfolds: Life at Mississippi State University

With the echoes of departure still resonating, Iyengar embarked on the next phase of his journey. The subsequent day saw him departing from Atlanta, bound for Columbus, Mississippi. This distance was approximately 26 miles, a mere thread of geography that separated Columbus from STARKVILLE – the cradle of the Mississippi State University (MSU) campus. As the wheels turned and the landscape shifted, a new chapter of his life was unfurling before him.

Image: Mississippi State University

Mississippi, situated in the "deep south" of the United States, bore the moniker of the Magnolia State, a tribute to the abundance of Magnolia trees that adorned its terrain. Its name was inextricably tied to the mighty Mississippi River, a majestic artery that originated in Minnesota, weaving its way through state borders before finding its ultimate embrace with the Gulf of Mexico. Yet, the Mississippi River held more than just geographical significance; it was a conduit of history, etching the narratives of settlers and enslaved Africans onto its banks.

As Iyengar stepped into Mississippi's embrace, he was greeted by a diverse demographic tapestry. Early settlers – French and Spanish explorers – had left their indelible mark alongside the footprints of the enslaved Africans who toiled upon plantations. The African American community constituted a significant portion of the population, reflecting the threads of history woven into the state's fabric.

The Mississippi River, with its meandering journey, cast a spell upon Iyengar. Its course brought to mind the Cauvery River, the lifeblood of Hemmige, his birthplace in India. In its waters, he found a familiar echo of nature's beauty, an ethereal connection that spanned continents. With each ripple of the river, he traced his journey – from the tranquil waters of his homeland to the banks of a foreign land.

Image: African American Community in Mississippi

Amidst the allure of the Mississippi River and the southern charm that permeated the air, Iyengar transitioned into his life at Mississippi State University. Here, academia intertwined with experience, shaping the contours of his narrative. The academic journey he had embarked upon mirrored the river's course – winding, unpredictable, yet teeming with life and potential.

In this newfound terrain, challenges and triumphs converged, becoming inseparable companions on his academic odyssey. The echoes of his past, the lessons of resilience he had learned, and the strength of his convictions coalesced to guide him through the uncharted waters of life at MSU. With the Magnolia State as his backdrop and the Mississippi River as his muse, Iyengar's journey continued, etching a unique chapter in the chronicles of his life.

Lessons Learned

"When shifts and transitions shake you to the core, see that as a sign of greatness that's about to occur."
-Russel Ballard

"The moment we believe that success is determined by an ingrained level of ability as opposed to resilience and hard work, we will be brittle in the face of adversity."
-Joshua Waitzkin

S.S. Iyengar knew adversity from the moment he set his sights on continuing his doctoral training in the United States. The transition of leaving family and friends for a new, unexplored area required determination and resilience. The shifts and transitions did shake young

---

Iyengar to his core, but he knew that his greatness lay beyond the path of his achievements in India.

He also knew that his resilience and hard work would be the sustaining factor in determining his success in this new adventure. Despite the adversities which lay ahead, Iyengar was resolved to work through the challenges. The excitement that coursed through his body on his way to America only served to reinforce his determination to succeed.

Chapter 3:
Life in Mississippi during the turbulent times

"Where there was despair I saw hope, where there was darkness I saw light, where there was sadness, I found joy. I spent from Sept 1970-December 1973 at MSU campus."

-S.S. Iyengar

3.0 The Transformative Decade: Navigating Turbulence in the 1970s

The 1970s emerged as a decade poised on the threshold of transformation, echoing the reverberations of the preceding 1960s. As the pages of history turned, the world found itself grappling with an array of pivotal events that would leave an indelible mark on society, culture, and politics. The 1970s, in many ways, became a bridge between the tumultuous past and an uncertain future. At the helm of political decisions was U.S. President Richard Nixon, who orchestrated an audacious incursion into Cambodia, expanding the already tumultuous Vietnam War. Amidst the backdrop of the war's escalating turmoil, the U.S. Senate made a profound move by revoking the Gulf of Tonkin resolution. This decision marked a critical juncture, relinquishing the expansive authority granted to Presidents Johnson and Nixon in the conduct of the war.

Even as the world grappled with the repercussions of war, the music scene underwent a seismic shift with the disbandment of The Beatles – a momentous event that mirrored the changing musical landscape. Beyond the realm of music, the political realm experienced a seismic shift with the passing of Egyptian President Gamal Abdel-Nasser, a figure whose influence stretched beyond borders. However, it was the fabric of American society itself that bore the imprint of the decade's turbulence. Violent clashes during student demonstrations, notably the tragic Kent State and Jackson State shootings, etched somber lines into the nation's narrative. In the very heart of the American consciousness, restlessness stirred, and the questioning of established authority became an undercurrent that permeated various aspects of society.

Image: Kent State and Jackson State Shootings

The 1970s became a canvas upon which self-expression flourished – a realm where public protests and individual fashion choices alike embodied the spirit of a generation seeking to carve its own path. The battle for equality persisted, with women, African Americans, Native Americans, LGBTQ+ individuals, and other marginalized groups pushing boundaries and demanding a more inclusive future. The echoes of the civil rights movement, the feminist movement, and the LGBTQ+ rights movement reverberated through the decade, leaving an indelible legacy.

Amidst the whirlwind of change, the 1970s emerged as a testing ground for the resilience of society – a terrain where the tempest of challenges met the crucible of change. As the world grappled with uncertainty, it was the resilience of the human spirit that stood as a testament to the indomitable will to forge a future that resonated with the cries of justice, the melodies of equality, and the harmony of progress.

3.1 Mississippi's Evolution: Navigating Past Struggles in the 1970s

In the autumn of 1970, as Iyengar set foot in Mississippi, he stepped into a state still wrestling with the reverberations of the American Civil War. Mississippi's history bore the weight of a bygone era, and its progress in educating the descendants of former slaves was a slow, cautious march. This was a state that had once led the nation in cotton production during the early 1800s; its prosperity was built on the foundations of expansive plantations and the labor of enslaved black individuals. However, the aftermath of the devastating Civil War in 1865 had left Mississippi culturally, intellectually, and economically devastated.

In the wake of the war, the period of Reconstruction brought forth significant challenges for Mississippi. The state, which had been among the first to secede from the Union to form the Confederate States of America, now grappled with the aftermath. With a newly emancipated black population accounting for over 55% of the state's total populace and a defeated white population contending with the stern reconstruction efforts of the federal government, progress became an uphill struggle.

For both black and white communities in Mississippi, the task of rebuilding their economy and forging new social structures was daunting. The United States Congress established the Freedmen's Bureau in 1865 to aid freed black individuals and white refugees by providing essential supplies. Despite encountering resistance and limited resources, the Bureau managed to establish some of Mississippi's initial public schools, planting the seeds for a future era of southern public education.

Image: Freedmen's Bureau, 1865

Yet, white Mississippians resisted federal intervention and the newfound freedom of black citizens. The emergence of the Black Codes mirrored the oppressive slave codes, imposing harsh restrictions on former slaves, now known as freedmen.

These codes governed everything from vagrancy to public gestures, aiming to curtail the liberties of black individuals. Economically, resistance brought about land price declines, the collapse of cotton markets, and near economic ruin. Over time, both black and white communities recognized the need for compromise, leading to the sharecropping system, where landowners provided resources and freedmen contributed labor, sharing the crop's yield.

Amidst this backdrop, the United States Congress embarked on a more comprehensive period of reconstruction from 1867 to 1876. This era witnessed substantial black participation in democracy, with over 226 black Mississippians holding public office – a remarkable feat compared to many other Southern states. However, this progress was hindered by prevailing sentiments of white supremacy, which facilitated the rise of groups like the Ku Klux Klan, perpetrating violence and racial terror. Formal reconstruction concluded in 1876, yet the uneasy coexistence between white and black communities persisted under Jim Crow laws, institutionalizing racial segregation. The 1896 Plessy v. Ferguson Supreme Court ruling upheld the "separate but equal" doctrine, further solidifying the division. By the time the 1970s arrived, Mississippi was undergoing transformations spurred by the Civil Rights movement and the shifting sociopolitical landscape. The state's education system was changing, reflecting a concerted effort to break free from the shadows of the past and usher in a new era of equality.

As Iyengar embarked on his journey in Mississippi, he found himself amidst a state navigating its intricate history, poised for renewal amid the winds of change and the pursuit of a more equitable future.

3.2 Navigating Uncharted Waters: A Newcomer's Journey in 1970s Mississippi

On the fateful day of September 12, 1970, Iyengar embarked on a transformative odyssey, leaving behind the familiar comforts of Bangalore, India, to embrace the humid embrace of Starkville, Mississippi, as a graduate student at MSU. This bold step symbolized a transition from the known to the unknown as he grappled with the challenges of a foreign culture, fortifying himself against obstacles and ultimately carving a path of remarkable success that would reverberate globally.

Upon his arrival at MSU's sprawling campus, Iyengar delved into the labyrinthine world of immigration formalities, a rite of passage for newcomers. His first port of call was his academic department, where he encountered Professor Carley. Awaiting him was a scholarship offer, accompanied by a stipend of 150 USD per year. His new role involved assisting Professor Jasper in the realm of Power Engineering. Under Dr. Jasper's guidance, Iyengar was directed to convene at the laboratory the following day for an in-depth discussion regarding the forthcoming tasks. The appointment was scheduled for the precise hour of 1:30 PM.

Yet, as life often unfolds, unforeseen circumstances emerge. Iyengar found himself delayed by 15 minutes in the crucial meeting. This slight tardiness, though seemingly inconsequential, led to an unexpected outcome – Professor Jasper's disapproval. The consequence was an impromptu lecture on the importance of punctuality. This incident left Iyengar wrestling with feelings of unease, harboring a fear that he had inaugurated his new chapter on an unfavorable note, thereby potentially compromising his credibility and rapport with his newly appointed superior.

As the sands of time continued to trickle, this initial hiccup seemed to cast a looming shadow over Iyengar's relationship with Professor Jasper. The professor's demeanor remained aloof and distant, causing Iyengar to tread cautiously through the ensuing semester. With the arrival of Spring in 1971, the culmination of his fears materialized – his assistantship under Professor Jasper was unexpectedly terminated.

Iyengar's journey through the early days in Mississippi offers a poignant illustration of the complexities and challenges that often confront first-generation immigrants as they strive for success in unfamiliar terrain. This narrative transcends mere chronology, encapsulating not only resilience in the face of adversity but also the essence of eventual triumph and global recognition for his profound contributions. It's a story of forging ahead amidst uncertainties, demonstrating how even the most tumultuous beginnings can set the stage for remarkable achievements.

3.3 Unveiling Purpose in Each Step: A Journey of Convergence

Amidst the tapestry of new opportunities, Iyengar's path serendipitously led him to Professor Bill Brown. This academic juncture bore the fruit of a collaborative project fusing Bio Engineering and Computer Science, offering Iyengar a monthly stipend of 300 USD. With seamless adaptability, Iyengar immersed himself in this role, quickly showcasing his intellectual acumen and prowess, much to the admiration of his peers.

As the months ebbed and flowed, Iyengar's unwavering dedication translated into financial stability. Saving a portion of his earnings, he sent remittances to his father in India, alleviating family expenses and contributing to the construction of a family residence in Bangalore. Beyond being a mere source of employment, Iyengar's collaboration with Dr. Brown metamorphosed into a profound catalyst that sparked his passion for the world of Computer Science.

Iyengar's innate mathematical aptitude and the realm of Computer Science seamlessly converged. This confluence of disciplines set the stage for his future trajectory. The summer of 1971 bore witness to Iyengar's multi-faceted endeavors, ranging from packaging televisions at Zenith Co. to his contributions at Holiday Inn. These experiences transcended routine employment; they were instrumental in shaping Iyengar's character, imparting a deep appreciation for the dignity of labor, a sentiment that would guide his choices in the years to come.

Image: Zenith Co., where Dr. Iyengar worked for earning

The synergy between high-performance computers and intricate biological system models was a transformative force. It paved the way for refined mathematical modeling in the realm of biological systems, propelling Computer Science into diverse applications. Iyengar's focus centered on cohesive data structure concepts intertwined with potent algorithms, a pursuit that enabled him to dissect complex biological systems with precision. His collaboration with Dr. Brown's lab bore fruit in the form of algorithmic and mathematical publications that garnered Dr. Brown's profound appreciation. However, fate's hand unveiled an unexpected twist. The summer of 1972 brought somber tidings from Albany, Georgia, as Iyengar received news of his grandmother's passing in Bangalore, India. Grief cast its pall over him for weeks, reminding him that sometimes the journey's end justifies the struggles faced along the way.

"My life started in September 12 (1970). Life on MSU campus. **I went to the department and met Prof. Carley and asked me to sign a contract for $150 dollars for a year.** *Job is to assist a Professor in Power Engg. His name is Dr. Jasper. He asked me to see mat 1:30pm the next day for discussion.* **I was late by 15 minutes. He gave a lecture the importance of good time sense."**

"I am afraid that I lost credibility with him with this incident. He lost all interest on my career and was avoiding him. **He cut my assistantship in Spring 1971."**

"I believe that everything happens for a reason"
" found a different person by name Bill Brown (jointly with Bio Engg. and Computer science) and **paid $ 300 dollars /month.** *The scholastic acumen fetched me with this amount. The money that I had saved sent to my father as he was trying to build a home in Bangalore, India. This is how I started shaping my interest in Computer science (as I was good in MATHEMATICS and*

> **"During the summer 1971, I did work in Zenith co. for packing television and also at Holiday Inn and other places. This summer job has helped me to shape my character and personality. I understood the dignity of labor in years to come"**
> **-S.S. Iyengar**

'Separate but Not Equal': A Lesson in Injustice on the Road to Equality

In the fall of 1972, during his second year as a doctoral student at Mississippi State University, Mr. S.S. Iyengar encountered a pivotal moment that challenged his understanding of justice and equality in America. Amid the rigorous demands of his PhD program, he faced an experience that would fundamentally reshape his views on race relations in the segregated South.

Image: S.S. Iyengar's 1972 Ford Galaxy

One afternoon, while driving his Ford Galaxy—a blue car he had purchased for $600—along Highway 82 near the Starkville campus, Iyengar became briefly distracted, resulting in an unfortunate accident with another vehicle. The car, driven by an African American, sustained considerable damage. Although the accident was undeniably Iyengar's fault, the aftermath revealed a deeper, more insidious layer of injustice.

When the police arrived, they assessed the situation and opted to issue tickets to both drivers instead of placing blame solely on Iyengar. This required both men to appear before a judge in Oktibbeha County, Starkville. Iyengar was struck by the equal treatment afforded to both drivers despite his clear responsibility for the crash. He understood that, while he faced primarily material consequences, the African American driver would bear far graver repercussions—not only the loss of his vehicle but also the burden of systemic discrimination that loomed heavily in the South.

Image: The Other Driver ('Pictorial Representation')

Days later, both men stood in court before a judge who held the power to determine their fates. Mr. Iyengar, acknowledging his mistake, expressed genuine remorse for the damage caused. The African American, in turn, outlined how the incident affected his life and livelihood. Yet, despite the clarity of Iyengar's fault, the judge abruptly dismissed the case without explanation, leaving both men perplexed.

For Iyengar, the judge's decision felt hollow, as if it trivialized the African American driver's struggles and failed to provide any real sense of justice. In that moment, he began to grasp the pervasive inequalities that defined the social and legal landscape of the time. While he returned to his studies with minimal lasting consequences, the African American driver would likely carry the compounded burdens of both the accident and the racial prejudice deeply rooted in Southern society.

This experience was transformative for Iyengar. He realized that although he could walk away relatively unscathed, the African American was subjected to structural inequities that denied him true justice. The incident illuminated the stark disparities between their experiences within the same system, marking the beginning of Iyengar's commitment to advocating for fairness and equality.

From that day forward, Iyengar carried the weight of that afternoon as both a personal lesson and a broader reminder of the work still required. His studies and professional pursuits increasingly reflected a desire to understand the societal structures that perpetuated inequality, compelling him to leverage his education and platform to challenge these injustices.

What began as an unfortunate mishap evolved into a catalyst for Iyengar's heightened sense of responsibility to advocate for marginalized communities. This incident spurred him to confront issues of race, privilege, and justice, influencing much of his later work as an educator and thought leader. In his journey, Iyengar came to understand that justice was not merely about legality; it was about ensuring that every individual—regardless of race—was treated with the dignity and fairness they deserved. Ultimately, the failure of the justice system to hold Iyengar accountable while the African American faced the brunt of the consequences underscored the enduring truth of "separate but not equal."

The subsequent months were marked by the pursuit of his dissertation. As August 1973 approached, a setback emerged when Dr. Brown revealed the unreliability of the collected data. This revelation left Iyengar disheartened. Despite being three years into his program, he ventured into new territory, exploring a fresh problem. Through convincing discussions, he proposed a statistical model to address the data gap. With Dr. Brown's endorsement, Iyengar continued to produce results, culminating in an exam scheduled for December 17th, 1973 – a pivotal milestone that propelled him forward.

> *"Things were going good and I published many papers both algorithmic and mathematical in nature. Dr. Brown was quite happy."*

"During the year 1972 close to summer I heard from my brother who was in Albany, Georgia that my Grandmother passed away in Bangalore, India,
This was quite painful and felt lonely for several weeks."

The end justifies the means

"The next 1 year was uneventful and during the summer of 1973, I was planning to start writing my dissertation and somewhere in August 1973 (Friday), Dr. Brown my data that I had collected during the summer is not good or very reliable. I was upset and he suggested that I should look for a different problem. On the same day in the evening myself and Subba Rao were planning for a ride to go to Gatlinberg, Tennessee"

*"After my discussion with Brown, I left perplexed not knowing what to do. After a week I had a meeting with him and I told him that will come up with a statistical modeling to compensate for missing data. I continued working on some more results and he asked me to have the **exam on December 17th** and after the exam I passed and never saw him. **He suggested to me that I should go back to India and he never helped me after that."***
-S.S. Iyengar

Upon conquering this milestone, his interactions with Dr. Brown faded. The mentor's advice to return to India was given with minimal support, a sentiment that could have been disheartening. Yet, Iyengar's tenacious spirit remained unbroken. Fueled by his determination to make a lasting impact on the U.S. academic landscape, fostering change for individuals and the nation, he transcended the indifference displayed by his former mentor.

His exceptional academic achievements, coupled with his influential contributions as a graduate student, served as the foundation for his role as an Assistant Professor at Jackson State University. The chapters that followed would recount Iyengar's life within the realm of a Historically Black University (HBCU), where the challenges of being an immigrant are interwoven with the narrative of transformation and growth.

> ***Lessons Learned***
>
> *"Education is not the learning of facts, but the training of the mind to think."*
> *-Albert Einstein*
>
> *"You may encounter many defeats, but you must not be defeated. In fact, it may be necessary to encounter the defeats so you can know who you are, what you can rise from, how you can still come out of it."*
> *-Maya Angelou*
>
> The pursuit of a doctoral degree is no trivial matter. Although there are many quotes that remind us of the doctoral journey, these two quotes perhaps best sum up the experience. Training the mind to think logically and critically, rather than remembering facts and figures, is the core of computer science and other science, technology, engineering, and math (STEM) disciplines.

In the laboratory setting, as in life, we may initially be assigned a mentor or boss and assume our work will be smooth sailing. However, a multitude of conflicts can arise, from personality conflicts to a shift of interest in the project or work. Iyengar encountered this with his first professor, which resulted in his leaving his position and the professor's lab.

Iyengar was also intertwined in the continuing undercurrent of racial segregation in Mississippi in the 70s. As an immigrant, Iyengar was unaware of this racial undercurrent. Like Rosa Parks before him, Iyengar experienced several incidents where he was told to "move to the back of the bus" on Mississippi's public transportation. It was only later that he understood the context of the request.

At the university, Iyengar discovered his true potential through his interactions with faculty members. With their guidance, he achieved significant success in his research and in publishing scientific papers. However, the extent to which racial or cultural prejudice influenced Iyengar's perception when he was advised to return to India by some individuals at the university remains unknown. It is important to consider that this occurred during a turbulent era in the Deep South.

It was through Iyengar's determination, resilience and understanding of who he was that enabled him to rise above the challenges, seek a faculty position in the United States at Jackson State College (now – Jackson State University), and help others in their life journey.

Chapter 4:
The Rise of the Circuit Rider

Iyengar's Philosophy - *"Education is the foundation for economic development and prosperity. It is the access point by which groups of all backgrounds are able to capture the American Dream."*

Over the past 50 years, he has lived that dream, opening doors for countless others while serving in three unique institutions.

"The phrase "Protect all without regard to complexion" underscores a commitment to equality and justice for all individuals, regardless of their skin color." — Dave Collins's Remarks, Oct 26. 2024 (AP News)

In 1831, New Haven, Connecticut, marked a pivotal moment in the rise of the abolitionist movement. This year witnessed the convening of a significant assembly of people of color in Philadelphia, which played a crucial role in advocating for the rights and freedoms of Black individuals. Additionally, during this period, the establishment of educational institutions for Black men was proposed and organized, reflecting a growing emphasis on education as a pathway to liberation.

This era was not only about addressing the immediate injustices of slavery but also about fostering a sense of community and interconnectedness among Black Americans. The efforts of this time laid the groundwork for lectures, discussions, and initiatives focused on Black liberation, symbolizing a collective awakening and the beginnings of organized resistance against oppression. Such movements were essential in shaping the early history of civil rights, igniting a passion for freedom and equality that would resonate for generations to come.

4.0 The Evolutionary Legacy of Jackson State University

Historically Black Colleges and Universities (HBCUs) play a crucial role in shaping the professional landscape for Black Americans. According to data released by the White House in 2022, HBCUs are responsible for producing an impressive 70% of Black physicians, 50% of Black lawyers, and 40% of Black engineers in the United States. This significant contribution underscores the importance of HBCUs in providing access to higher education and professional opportunities for Black students, fostering a diverse workforce that enriches various fields. By equipping students with the knowledge and skills needed to excel in these professions, HBCUs are vital in advancing social equity and representation in sectors that have historically been underrepresented.

The records of Jackson State University, a beacon of educational excellence, trace their origins to October 23, 1877. This storied institution, initially known as Natchez Seminary, emerged as a private school under the auspices of the American Baptist Home Mission Society of New York. Founded with a noble purpose, it sought to illuminate the path of education for the newly emancipated slaves of Mississippi. Through the passage of time, Jackson State University has undergone multiple metamorphoses, each marking a chapter of growth and progress.

In its nascent form, the institution was christened Natchez Seminary in 1877, nestled in Natchez, Mississippi. Its mission was to nurture Christian leaders within the African American populace of Mississippi and its neighboring states. However, the year 1882 heralded a pivotal shift as the campus found its new home in Jackson. With this transition came an expansion of the curriculum's scope, prompting a renaming to Jackson College in March 1899.

The dawn of the 1940s ushered in a new era as the state embraced stewardship of the college, defining its purpose as the cultivation of educators. The following decade, from 1953 to 1956, witnessed a remarkable surge of growth. During this transformative phase, the institution introduced a graduate program and launched bachelor's programs in arts and sciences. Reflecting these transformative strides, the name underwent another evolution, becoming Jackson State College in 1956.

The momentum of progress remained undeterred, accompanied by extensive curriculum expansion and infrastructural development. This crescendo reached its zenith on March 15, 1974, when Jackson State College achieved the revered status of a university. A few years later, in 1979, the institution was bestowed the prestigious title of the Urban University of the State of Mississippi.

Image: JSU in 1974

In the present epoch, Jackson State University stands as a publicly funded coeducational institution. Its financial sustenance is derived from legislative allocations, supplemented by student fees, private grants, and federal funding. Guided by an enduring mission to empower a diverse student body to emerge as leaders, the university aspires to be celebrated as an intellectually stimulating and technologically advanced intellectual enclave. Within this dynamic and vibrant milieu, students and faculty seamlessly weave innovative research, engage in interdisciplinary collaborations, and contribute to the global community, emblematic of the indomitable spirit that has propelled Jackson State University's journey through time.

4.1 Iyengar's Transformational Journey at JSU: Pioneering a New Era in Computer Science Education

The year 1974 marked a watershed moment in the life of Dr. S.S. Iyengar as he embarked on an academic odyssey that would indelibly reshape the landscape of computer science education. Jackson State University (JSU), with its rich history and commitment to fostering academic excellence, became the canvas on which Dr. Iyengar painted his visionary endeavors. The genesis of this transformative chapter was ignited by an invitation extended by Dr. Jessi Lewis, a trailblazer in the world of education. Dr. Iyengar's expertise was sought to spearhead the development of a regional educational computing network, a forward-looking initiative funded by none other than the esteemed National Science Foundation (NSF). This network's audacious mission was to interconnect colleges across the state of Mississippi, effectively weaving together disparate educational institutions into a cohesive fabric powered by the immense potential of computing technology.

Dr. John Peoples Jr. (Former President of JSU) and Dr. Jesse Cornelius Lewis (Former Dean of Computer Services at JSU)

Dr. John Peoples Jr., the visionary president of Jackson State University (JSU), played an instrumental role in shaping the future of education for disadvantaged students in Mississippi. Sixty years ago, at a time when the state lacked robust support for minority education, particularly in the realm of technology, Dr. Peoples recognized the transformative potential of computer science education. His strategic vision was groundbreaking and forward-thinking, especially in a region and era where racial and economic inequities were deeply entrenched. He prioritized the integration of computer science into the university's curriculum, understanding that technological literacy would be key to breaking cycles of poverty and marginalization for many African American students.

At the time, Mississippi faced significant challenges in providing quality education to students from disadvantaged backgrounds. Funding for minority schools was limited, and the state's education system was largely segregated, with stark differences in resources between white and Black institutions. Despite these obstacles, Dr. Peoples saw an opportunity to leverage emerging technologies to empower students who had historically been excluded from the field of computer science. He believed that equipping these students with technical skills would not only enhance their future career prospects but also enable them to contribute to the growing national demand for technology professionals. His vision went beyond providing access to basic education—it was about creating pathways for economic and social mobility through cutting-edge technology.

Dr. Peoples' foresight was especially critical in a state that, during those years, struggled with educational inequities and lacked a concerted effort to uplift minority populations. His leadership at JSU ensured that computer science became a priority, not just an optional field of study. By emphasizing computer science education, he aimed to close the gap between minority and white students in access to technology and higher education opportunities. This initiative became a beacon of hope for many disadvantaged students who had previously been denied access to such opportunities.

In partnership with organizations like the National Science Foundation (NSF) and through collaboration with key educators such as Dr. S.S. Iyengar, Dr. Peoples laid the groundwork for transformative educational reform. His commitment to providing computer science education to underrepresented students set a precedent for other institutions in the state and region, advocating for a more inclusive and equitable approach to higher education. Dr. Peoples' vision was not just about keeping up with technological advancements but about ensuring that minority students were at the forefront of these developments, ready to lead and innovate.

His legacy in this regard continues to inspire efforts to enhance educational access for marginalized communities. Dr. Peoples' strategic prioritization of computer science at JSU helped position the university as a leader in technology education in Mississippi, creating lasting impacts that continue to resonate in today's digital world.

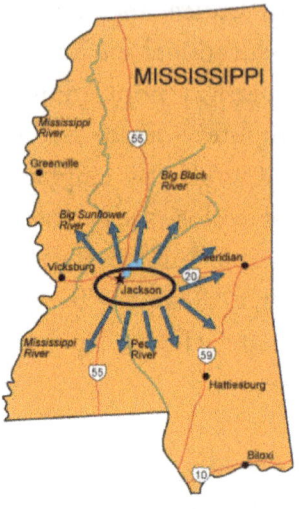

From June 1973 to 1979, the National Science Foundation (NSF) Education Network funded a transformative initiative aimed at addressing educational disparities in computer science education across Mississippi. The project was an ambitious and groundbreaking effort, designed to empower minority schools and underserved communities with advanced computational tools and resources, helping bridge the digital divide at a time when technology was rapidly evolving. The NSF provided essential funding to support the development of virtual educational tools, with a focus on making programming and computer science more accessible and understandable to students who had historically been excluded from these fields.

One of the key elements of this initiative was the provision of Texas Instruments interactive terminals, a revolutionary technology at the time, which allowed students to engage with programming in ways that were previously impossible. These interactive terminals provided hands-on experience, helping students grasp the practical aspects of computing while nurturing their problem-solving skills. This equipment significantly contributed to reshaping how computer science was taught in minority schools, helping students become familiar with the tools of the future. Beyond just offering resources, the NSF project made a significant impact on the accessibility of computer science education, setting the stage for future technological advances.

Dr. S.S. Iyengar, along with Dr. Pramanik, played a pivotal role in this project as circuit riders, a term used for educators and experts who traveled to various institutions to conduct workshops, training sessions, and provide technical assistance. These circuit riders visited both minority and predominantly white schools across Mississippi, ensuring that the project reached a broad range of institutions, regardless of their demographic makeup. Dr. Iyengar's commitment to educational equity was apparent in every facet of his involvement. He was deeply dedicated to creating educational systems that catered to a wide range of learning styles, skill levels, and institutional needs, ensuring that no student or school was left behind.

The impact of this initiative was widespread and far-reaching. Among the many institutions that benefited from this NSF-funded effort were Jackson State University, the host institution for the project, and several other historically Black institutions, including Alcorn State University, Tougaloo College, Rust College, and Mississippi Valley State University. These schools, many of which had limited resources for technological education, were given access to computational tools that would have otherwise been unattainable.

In addition to historically Black institutions, the project also reached predominantly white schools, demonstrating its inclusive approach to educational reform. Institutions such as Millsaps College, Mississippi College, Hinds Junior College, Gulf Coast Community College, and Belhaven College also participated in the initiative. This broad range of participating schools reflected a commitment to fostering collaboration across different types of institutions, bridging gaps not only in resources but also in cultural and academic divides. By fostering cross-institutional learning opportunities, the project helped to break down barriers between minority and white students, creating a more inclusive environment for all participants.

Workshops led by Dr. Iyengar and Dr. Pramanik were central to the project's success. These workshops, hosted at various institutions, provided students and faculty alike with training on the use of the Texas Instruments interactive terminals, as well as broader lessons in computational thinking and programming. The workshops also provided opportunities for collaboration between minority and white institutions, with participants sharing knowledge and resources in ways that were unprecedented at the time. The collaborative nature of these workshops fostered a spirit of mutual learning and respect among participants, contributing to long-lasting relationships between the institutions involved. More importantly, there was a strong support from IBM faculty exchange program to minority schools in addition to Bell Laboratories. These were very crucial for the success of this computer educational network.

Jackson State University, as the host institution, played a central role in coordinating these workshops and ensuring the smooth implementation of the program. Its leadership in the project was crucial in mobilizing resources and ensuring that minority institutions in Mississippi had access to the same level of technological education as their white counterparts.

Through the tireless efforts of Dr. Iyengar, Dr. Pramanik, and other leaders involved in the project, Mississippi's educational landscape began to shift, with computer science becoming more integrated into the curricula of schools that had long been left behind in technological advancements.

The success of this project was not only measured in terms of the number of institutions it reached but also in its long-term impact on computer science education in Mississippi. By equipping schools with the necessary tools and expertise, the NSF-funded initiative laid the foundation for future advancements in technology education, ensuring that minority students had a pathway to succeed in the rapidly growing field of computer science. Dr. Iyengar's unwavering commitment to this cause remains a testament to his dedication to educational equity, and the success of the project continues to inspire efforts to create more inclusive, accessible educational systems today.

Over the next five years, Dr. Iyengar donned multiple hats, assuming roles that ranged from network coordinator to Assistant Professor. He crisscrossed the expanse of Mississippi, traversing its towns and campuses, in his unwavering dedication to unite the state's educational institutions through the power of technology. His efforts encompassed not only the technical intricacies of programming but also the art of inspiring individuals with the profound impact that computer science could bring to their lives and their communities.

Yet, Dr. Iyengar's vision went beyond the technical realm and was grounded in a deep-seated sense of responsibility and equity. He strongly believed that computer science education should be accessible to everyone, regardless of their background. This conviction led him to conduct a series of computer science workshops in diverse locations, which were more than just technical sessions. These workshops served as platforms for empowerment, enlightenment, and the democratization of knowledge.

During his tenure at JSU, Dr. Iyengar orchestrated a symphony of collaboration, merging the realms of computer science with other disciplines. This interdisciplinary approach culminated in the conceptualization of a visionary Master's Degree graduate program in computer science. Such initiatives were rare in those early days, but Dr. Iyengar's foresight and tenacity prevailed. The program, like a phoenix rising from the ashes, heralded a new era for JSU's computer science department in the late 1970s. Dr. Shakti Pramanik, a fellow coordinator and a holder of a PhD in Computer Science from Yale, was a co-architect of this monumental endeavor.

As Dr. Iyengar's vision gained momentum, he left an indelible imprint through a diverse array of workshops and courses, making the world of computer science accessible to a widening spectrum of students. The Master's Degree program he pioneered grew into an intellectual powerhouse, imparting advanced computer science education to over 5,000 minority students. It's a testament to the program's quality that an astounding 25% of the inaugural cohort went on to earn their PhDs from prestigious institutions such as Iowa State, University of Maryland, and Purdue. Even more remarkable, over 60% of these graduates secured positions in renowned enterprises like Lockheed Martin and Raytheon, becoming the bedrock of innovation within these Fortune 500 corporations.

This program, which began as an idea, not only shaped individual careers but also sowed the seeds for a PhD program in Computational and Data Science. The legacy of Dr. Iyengar's commitment lives on as a beacon of inspiration and impact. His vision, his dedication, and his unfaltering commitment to advancing computer science education, particularly among underrepresented groups, continue to reverberate through the corridors of academia and industry. In retrospect, Dr. Iyengar's tenure at JSU stands as an era where pioneering spirit, relentless dedication, and visionary leadership converged to sculpt a new paradigm in computer science education, forever altering the trajectory of countless lives.

4.1.1 Inception of a Groundbreaking Graduate Program at JSU

In the crucible of the 1970s, Dr. S.S. Iyengar stood as a beacon of determination, undeterred by the arduous challenges faced by minority faculty in the realm of academia. The landscape of education during that era was rife with inequities and a scarcity of supportive environments that nurtured the growth and accomplishments of underrepresented educators. It was within this turbulent milieu that Dr. Iyengar embarked on a mission that would forever reshape the narrative of Jackson State University (JSU) and the future of computer science education.

Driven by a resolute desire for change, Iyengar seized a transformative idea: the initiation of a groundbreaking graduate program at JSU, offering both Master's (MS) and Doctorate (Ph.D.) degrees. Recognizing the dire need for more inclusive and empowering opportunities for minority students, especially in the South, Iyengar took it upon himself to bring this vision to fruition. With unwavering determination, he embarked on a journey that would challenge norms and reshape the trajectory of higher education.

The inception of this audacious proposition was met with skepticism, reflecting the prevailing sentiments of the time. However, Iyengar remained steadfast, his unwavering determination serving as a driving force. Through numerous rounds of discussions and persuasive dialogues, he engaged with the Dean and Head of the Computer Science department at JSU, advocating for the profound impact that this program could have on shaping the careers and successes of minority students and faculty members.

"During the 1970's when i was an Assistant professor at Jackson State university (1976) there was no supportive and equitable environment for faculty and initiatives to advance minority faculty careers and success. It is in this context that I suggested the Dean and Head of the CSC dept to start a new graduate program (MS and Ph.D)."

"Dr. Walker came and gave an overview of needs to be done to advance a graduate program but said lot of negative things about the quality of over all education program at JSU."

> *"Dr. Manning was sitting with me and sort of pushing me with his leg under the table to respond to his criticisms."*
>
> "After the meeting Dr. Manning told me that *we cannot talk directly to white folks and we always are not confident because in this part of the country we were not treated equal*. Probably you do not understand coming from a foreign country."
>
> *-S.S. Iyengar*

In a pivotal turn of events, JSU invited Dr. Terry Walker, Chair of the Computer Science department at the University of Southern Louisiana (USL), to assess and provide feedback on the proposed graduate program. Dr. Walker's expertise and role as a gatekeeper of computer science education lent weight to his evaluation. At that time, USL was a solitary bastion in the Southern United States offering a Ph.D. program in Computer Science.

Dr. Walker's assessment was accompanied by a list of prerequisites for the proposed program. While his feedback contained constructive elements, it carried undertones of skepticism regarding the quality of education at JSU. Despite the critique, Dr. Iyengar's unwavering determination propelled him to respond with a compelling argument that showcased JSU's readiness to embrace graduate programs and the immense benefits these programs could bring to the community, especially in empowering minority faculty members.

In a pivotal moment during the assessment meeting, Dr. Manning, seated alongside Iyengar, discreetly nudged him under the table. This subtle gesture served as a catalyst, urging Iyengar to take the lead in addressing Dr. Walker's concerns. And take the lead he did, presenting a formidable rebuttal that countered doubts about JSU's capabilities and highlighted the transformative potential of the proposed program.

As the meeting concluded, Iyengar sought clarity from Dr. Manning about his role in responding to Dr. Walker. In response, he was confronted with a stark truth that illuminated the stark realities of the educational landscape for minorities in the South. Dr. Manning revealed the hesitance to directly engage with white individuals due to the pervasive disparities that minorities encountered in this region. This revelation underscored the systemic challenges that underrepresented faculty faced in their quest to ascend within academia.

Image: JSU in 1977

In the face of adversity and against the backdrop of systemic barriers, Iyengar's persistence and determination triumphed. His relentless advocacy led to the ultimate approval of the Master's Program at JSU, a historic achievement that marked a profound turning point in his professional journey. As he assumed his role as a circuit rider/network coordinator at JSU with an annual salary of $11,000 USD, Iyengar's accomplishment resonated far beyond mere academia. It served as a testament to his unwavering commitment to expand opportunities for minorities in higher education, while blazing a trail for others to follow. The inception of this groundbreaking graduate program wasn't just an achievement for Iyengar, but a milestone that resonated as a triumph against adversity and a beacon of hope for generations to come.

4.2 Empowering Minority Universities: Challenges, Reflections, and Transformative Leadership

The landscape of education in the United States has been marked by a rich tapestry of institutions, each contributing to the complex mosaic of learning opportunities. Among these, Historically Black Colleges and Universities (HBCUs) emerged in the 19th century with the noble aim of providing education to African American students who were denied access to mainstream institutions due to systemic racism. However, while HBCUs played a pivotal role in breaking barriers and nurturing generations of leaders, they faced unique challenges that have persisted over time.

Image: Various HBCU's

The establishment of HBCUs aimed to create nurturing environments for African American students, but inadvertently created pockets of segregated education. This separation, while born out of necessity during an era of profound racial inequality, had unintended consequences. Attending an HBCU could limit students' exposure to diverse settings, potentially constraining their career prospects and their ability to navigate a multicultural world. Dr. Iyengar's own journey illuminated this limitation, as he recognized that HBCU attendees might miss out on valuable experiences that come from interacting with a broader range of perspectives.

Throughout his engagement with minority universities, Iyengar discerned common challenges that hindered their ability to compete on equal footing with their mainstream counterparts. These challenges included the lack of state-of-the-art facilities, limited student motivation, and a shortage of effective mentors and role models. Additionally, the confined campus environments and subpar infrastructure further compounded the hurdles these institutions faced. A deficiency in ambition to excel in various aspects also hindered the growth of students and faculty alike.

Beyond the internal challenges, Iyengar's experiences as a representative of a minority university exposed him to the harsh reality of discrimination. He encountered skepticism and unwarranted questions about his language skills due to his brown skin. These prejudiced encounters left him feeling marginalized and excluded from the broader educational landscape, highlighting the deeply entrenched biases that persist even within academia.

In the face of these complex challenges, Iyengar engaged in profound reflection, seeking ways to empower minority universities and foster their growth. He arrived at a powerful insight: transformative leadership rooted in global thinking is a pivotal element in overcoming these hurdles. This perspective became a guiding principle throughout his journey as an educator and administrator. He recognized that to truly enable minority institutions to thrive, a leadership approach that embraces diverse perspectives, fosters innovation, and transcends the confines of immediate surroundings is essential.

Iyengar's philosophy goes beyond addressing the challenges at the surface level. It involves instilling a sense of confidence and ambition within students and faculty, propelling them to break barriers and exceed expectations. It involves building robust mentorship networks that provide unwavering support and guidance, ensuring that talent is nurtured to its fullest potential. It involves creating an environment that encourages interdisciplinary collaboration, infusing global perspectives into the fabric of education.

In his role as an educator and administrator, Iyengar has consistently championed these principles. He has advocated for the modernization of facilities, the cultivation of mentorship programs, and the infusion of global perspectives into curricula. His commitment to transforming minority universities into thriving centers of excellence serves as a testament to his dedication to positive change. Through his visionary leadership, Iyengar aims to erase the limitations that have hindered the growth of minority institutions, enabling them to shine as beacons of learning, innovation, and empowerment.

4.2.1 Navigating Immigration: The Role of Legal Assistance

Iyengar's ambitious journey from India to the United States was not without its share of legal intricacies and challenges. His immigration pathway was intricately guided by the expertise of Mr. David Carliner, a prominent figure in the realms of immigration, civil liberties, and civil rights law, stationed in Washington, D.C. Carliner's profound understanding of American immigration and naturalization law played a pivotal role in ensuring Iyengar's successful immigration journey, which spanned over two years of collaborative effort.

Carliner's legal acumen extended beyond a mere adherence to statutes; it encapsulated a commitment to championing equality and justice. His advocacy extended to confronting anti-miscegenation laws and challenging the segregation of public accommodations. Such dedication to civil rights mirrored Iyengar's own aspirations for equality and opportunity.

As Iyengar pursued his immigration objectives, he incurred expenses amounting to approximately 2500 USD for Carliner's legal representation. This investment was more than financial; it symbolized an alignment of values and aspirations. Carliner's unique and distinctive approach to legal practice left an indelible mark on Iyengar's immigration journey, guiding him through a complex legal landscape with a personalized touch.

However, the initial phase of Iyengar's immigration journey encountered a roadblock when his application was rejected. This setback prompted a moment of reflection and consideration, leading Iyengar to explore alternative avenues. In this pivotal juncture, the idea of pursuing Canadian immigration emerged as a viable option. At the time, Canadian visas presented a relatively straightforward process compared to their American counterparts. This phase of uncertainty compelled Iyengar to evaluate his options and make strategic decisions. The contemplation of Canadian immigration provided him with an alternative perspective, allowing him to weigh the possibilities before ultimately reaffirming his commitment to the United States.

Iyengar's immigration journey, closely guided by Carliner's legal expertise, exemplified the intricate and multifaceted nature of international mobility. Beyond legal documentation, this journey mirrored the pursuit of dreams, the navigation of challenges, and the resilience required to overcome setbacks. As Iyengar's path unfolded, it underscored the significance of legal advocacy in shaping individual destinies and the broader narratives of equality and justice.

4.3 Empowering Through Circuit Riding: Iyengar's Impact

Hired under the auspices of the JSU NSF Educational Network, Iyengar assumed the role of a circuit rider, embarking on a mission to reshape computer science education in Mississippi. This transformative endeavor was fueled by the visionary leadership of Dr. Jesse Lewis, the head of the Computer Science department at JSU, and Dr. John A Peoples Jr, the President of JSU. Iyengar's circuit rider role encompassed a two-fold purpose that would reverberate through the corridors of education and opportunity:

- **Empowering Economically Disadvantaged Minority Students**

Iyengar's first mandate as a circuit rider was to provide academically gifted yet economically challenged minority students with a challenging educational opportunity. His commitment to this cause was a testament to his unwavering belief that talent knows no socio-economic boundaries. By nurturing these students' potential, Iyengar aimed to uplift them from their challenging circumstances and equip them with the tools to succeed academically and professionally.

- **Extending Quality Education to Underserved Communities**

One of the defining aspects of Iyengar's journey as a circuit rider was his unwavering dedication to bridging the educational gap that had been caused by systemic racism and socio-economic disparities. He recognized that many students of color, especially those living in urban areas across the United States, were being denied access to quality educational programs due to deeply ingrained biases and financial constraints.

As a circuit rider, Iyengar made it his mission to dismantle these barriers and ensure that all students, no matter their background, could access world-class computer science education. He believed that education was the key to unlocking a brighter future for these students and empowering them to reach their full potential.

Throughout his journey, Iyengar achieved several significant milestones that serve as a testament to his unrelenting commitment to education and empowerment. His legacy continues to inspire and motivate educators and students alike, as the world strives towards a more equitable and inclusive society.

- **Creation of a New Master's Program**

A landmark achievement during Iyengar's tenure was the establishment of a pioneering Masters program in computer science at Jackson State University. This program was designed to provide students with a transformative learning experience, guided by the ethos of academic excellence. An exemplar of this initiative was the course titled "Data Structures and File Management," offered at the graduate level. The course not only imparted technical skills but also fostered critical thinking and problem-solving abilities. Its impact resonated far beyond the classroom, contributing to the holistic development of students.

For Details: https://dl.acm.org/doi/10.1145/382222.382458

- **Crafting a Unique Graduate Program**

In 1976, Iyengar's visionary leadership gave rise to a distinctive graduate program in computer science at Jackson State University. This initiative was not merely about disseminating knowledge; it was about empowering students to become trailblazers in the field. The program encapsulated the essence of Iyengar's commitment to equity and excellence. Its impact transcended institutional boundaries, influencing the trajectory of numerous lives and inspiring a new generation of leaders.

For Details:
https://dl.acm.org/doi/abs/10.1145/953026.803498

A unique graduate program in computer science at Jackson State University

Author: S. Sitharama Iyengar Chairman: Jesse C. Lewis   Authors Info & Claims

ACM SIGCUE Outlook, Volume 10, Issue 3 • Pages 355 - 358 • https://doi.org/10.1145/953026.803458

Published: 01 February 1976 Publication History   Check for updates

Abstract

The two key elements of decision-making in today's environment are (i) efficient information processing and management and (ii) a degree of maturity in effective use of a broad range of analytical tools and techniques which pass under the general label of Mathematical Sciences.

The Department of Computer Science at Jackson State University started a Graduate Program during the 1974 academic year which offers a unique interdisciplinary program leading to a Master's Degree in Computer Science. This program emphasizes in programming languages, systems programming, operating systems, information systems analysis and design, the role of the computer as an integral part of the decision-making process, and the computer applications in the areas of statistics and management science. Advanced placement may be given via examination or acceptable certificate presentation from traditional or non-traditional institutions such as IBM Corporation. The students are presently using both the interactive and the batch capability of our IBM 360/40 for their course and project works. The university is upgrading to an IBM 370/145 during the spring semester (1976).

4.4 Expanding Horizons: The Circuit Rider's Journey in Enhancing Minority Education

Iyengar's transformative role as a circuit rider extended beyond the confines of Jackson State University, reaching out to enhance the education of minorities across the United States. Driven by a deep understanding of the critical need for robust programming skills in the digital age, Iyengar embarked on a mission to empower minority students with these essential abilities. Collaborating with esteemed colleagues such as Dr. Vijay Asher and Dr. Lynn Bracker, he harnessed his expertise to create a lasting impact.

The catalyst for this endeavor was the recognition that strong programming skills held the potential to open doors of opportunity for minority students, equipping them with a competitive edge in an evolving technological landscape. Armed with this conviction, Iyengar sought support from the IBM faculty loan program, a remarkable initiative designed to channel the expertise of volunteers from IBM towards assisting engineering schools' minority education programs. This program served as a beacon of empowerment, offering financial backing and resources to drive change. Guided by the core principles of this program, Iyengar and his dedicated colleagues embarked on an ambitious journey that spanned numerous educational institutions. Their mission was twofold:

- **Empower the Academically Talented, Economically Disadvantaged**

A cornerstone of Iyengar's efforts was to provide academically talented yet economically disadvantaged minority students with an avenue to succeed. He recognized the potential within these students and was determined to provide them with the tools to flourish in their academic pursuits.

- **Create Challenging Educational Opportunities**

At the heart of Iyengar's vision was the belief that educational opportunities should challenge and inspire students. Armed with this ethos, he set out to craft workshops that would not only impart knowledge but also foster critical thinking and problem-solving skills.

To achieve these goals, Iyengar and his collaborators designed programming workshops centered around BASIC (Beginners All-Purpose Symbolic Instruction Code), a language known for its accessibility and ability to demystify programming logic. Their workshops spanned two to three days and were hosted by institutions such as Mississippi College under the guidance of Dr. Paul Ohme, a visionary educator himself. These workshops weren't just about imparting knowledge; they were about nurturing a passion for programming and equipping students with practical skills.

The impact of Iyengar's efforts extended far beyond the workshop sessions. At the culmination of each workshop, the team donated TEXAS INSTRUMENT TERMINALS along with essential software to the participating colleges. This hands-on equipment empowered students to put their newfound skills into practice, bridging the gap between theory and application.

Iyengar's journey in enhancing minority education through programming workshops stands as a testament to his commitment to equity and empowerment. His initiatives didn't just impart technical skills; they ignited a spark of curiosity and ambition within students, guiding them towards a path of meaningful contribution and success. Through his dedicated efforts, Iyengar ensured that the doors of opportunity swung wide open for aspiring minds, transcending barriers and shaping the future of minority education.

4.4.1 The Multifaceted Goals of Empowerment

Iyengar's commitment to enhancing minority education went beyond the confines of programming workshops; it encompassed a multifaceted approach aimed at comprehensive transformation. The IBM faculty loan program not only provided financial support but also laid the foundation for a range of initiatives that aimed to empower both students and institutions. These initiatives collectively underscored Iyengar's unwavering dedication to fostering equitable and inclusive educational environments.

1.   **Facilitating Policy and Infrastructure Discussions:** Central to Iyengar's mission was the recognition that effective integration of computer technology required well-defined policies and a robust infrastructure. He engaged in discussions about policies, hardware, software, and course offerings that would facilitate students' utilization of computers as effective tools for learning.

2.   **Developing Comprehensive Course Guidelines:** Recognizing the need for standardized education, Iyengar and his team worked diligently to create course guidelines and requirements for both undergraduate and graduate students. These guidelines were designed to ensure consistency and quality in computer science education across institutions.

3.   **Training Faculty and Students:** To bridge the knowledge gap, Iyengar took on the task of training both faculty and students in the effective use of computer terminals connected to the network. By providing hands-on training, he aimed to empower educators and learners with the skills needed to leverage technology for enhanced learning experiences.

4.   **Leadership in Curriculum Development:** Iyengar's leadership extended to collaborative efforts with Jackson State University in shaping regional networks and workshops. Through these joint initiatives, he sought to expand the reach of quality computer science education to a broader spectrum of students.

5.   **Sharing Experiences and Insights:** One of the key facets of Iyengar's role was to provide valuable feedback and reports on how other institutions could

replicate Jackson State University's successful computing experiences. He served as a guiding light for institutions seeking to emulate effective practices.

6. **Assessing Instructional Computing:** Iyengar undertook the significant task of assessing the instructional computing landscape. He built a model that drew insights from a diverse range of institutions, including secondary schools, public school districts, community colleges, colleges, and universities across the United States. This endeavor facilitated the identification of best practices and optimal approaches.

7. **Enhancing Computer Center Support:** Acknowledging the pivotal role of computer centers, Iyengar worked to organize and staff these centers to provide improved support for instructional computing activities. This proactive step aimed to ensure that students and faculty could make the most of available resources.

8. **Fostering Computer Literacy:** Understanding the importance of computer literacy, Iyengar's efforts extended to initiatives aimed at raising the general level of computer literacy on campuses. By educating students and faculty about the applications and societal impact of computers, he empowered them to navigate the digital landscape effectively.

9. **Expanding Computer Science Curricula:** Iyengar recognized the value of a robust computer science curriculum. He played a role in establishing and improving these curricula, ensuring that students received comprehensive training in computer science and data processing.

Through his multifaceted initiatives, Iyengar's impact reverberated across institutions, students, and faculties, making strides towards equitable education and empowered learning. His tireless dedication to enhancing minority education laid the groundwork for a more inclusive and technologically empowered educational landscape.

4.4.2 Empowering Through Collaboration: A Network of Excellence

Dr. S.S. Iyengar's transformative work as a circuit rider was not confined to the walls of a single institution; instead, it was a broad-reaching mission aimed at enhancing education in computing and programming concepts for students and educators across a variety of institutions. Through his collaborations with universities of diverse backgrounds—both historically black and historically white institutions—he pursued his vision of broadening educational horizons and promoting equity in the computing field.

A pivotal figure in this endeavor was Dr. Jesse Cornelius Lewis, Dean of Computer Services at Jackson State University (JSU). Dr. Lewis played a crucial role in Iyengar's early career by offering him his first job as a circuit rider in a minority-serving institution in 1974. With a salary of $11,000 per year, Iyengar took on the task of establishing and directing a statewide Educational Computing Network that provided vital computer services to nineteen other institutions across the state. This network became a foundational element in Iyengar's broader strategy of using technology to bridge educational gaps between underrepresented and well-resourced institutions. His work at JSU also included the establishment of B.S. and M.S.

programs in Computer Science, creating lasting educational pathways for future generations of students at JSU.

Dr. Iyengar's work with the Educational Computing Network at Jackson State University exemplifies his commitment to democratizing education through technology. By establishing a system that allowed nineteen institutions, many of which had limited resources, to access cutting-edge computer services, Iyengar ensured that students at these institutions had the tools and opportunities needed to excel in a rapidly changing technological landscape. The creation of the B.S. and M.S. programs in Computer Science at JSU further solidified his impact, providing a structured and comprehensive path for students to gain deep knowledge in the field of computing.

In addition to his direct impact on institutions like Jackson State University, Dr. Iyengar's work as a circuit rider had a ripple effect across the country. His collaborations with institutions in both majority and minority communities reflected a dedication to diversity and inclusivity. The NSF Education Network initiative provided the platform for these collaborations, enabling universities to work together on creating innovative solutions for educating the next generation of computer scientists.

Dr. Jesse Lewis, Iyengar's collaborator and mentor during this period, had an equally impressive career trajectory. From 1984 to 1997, Dr. Lewis served as Vice President for Academic Affairs at Norfolk State University in Norfolk, Virginia. In this role, he continued his commitment to advancing higher education and expanding opportunities for minority students. His professional relationships extended beyond academia, as he worked closely with organizations such as People-to-People, the United States Agency for International

Development (USAID), and the NSF. Upon his retirement, Dr. Lewis was recognized as Senior Scientist Emeritus, a testament to his long-lasting contributions to the fields of education and computing.

Dr. Iyengar's early experiences working with Dr. Lewis and other visionary leaders shaped his broader philosophy of empowering through collaboration. His focus on creating a network of excellence, driven by the belief that collaboration between institutions is key to overcoming educational disparities, became a hallmark of his career. By connecting historically underrepresented institutions with cutting-edge computing technologies and educational tools, Iyengar was able to create pathways to success for students who otherwise might not have had access to these resources.

The legacy of this collaborative effort continues to influence educational strategies in computer science today. As the demand for highly skilled professionals in the field grows, the importance of creating equitable access to education in computing has never been more critical. Dr. Iyengar's early work, supported by the NSF and guided by leaders like Dr. Lewis, laid the groundwork for a more inclusive and dynamic future in the field of computer science education.

In sum, Dr. S.S. Iyengar's collaborative work through the NSF Education Network and his leadership at Jackson State University exemplify a visionary approach to broadening access to computer science education. His initiatives helped build the foundation for equitable opportunities in the field, ensuring that students from underrepresented institutions were equipped to succeed in the rapidly evolving technological landscape. His enduring contributions to the field of computer science education and his commitment to diversity and inclusion have left a lasting legacy, empowering future generations to thrive in technology-driven fields.

The list of universities where Iyengar's influence was felt is an embodiment of his dedication to fostering inclusivity and excellence across institutions:

- **Jackson State University (Host Historically Black Institution) (Founded 1877)**

Jackson State University traces its origins to Natchez Seminary, established on October 23, 1877, in Natchez, Mississippi. The seminary was founded by the American Baptist Home Mission Society of New York with the mission of fostering the moral, religious, and intellectual growth of Christian leaders among Mississippi's Black population and those in neighboring states. In 1883, the school was renamed Jackson College and relocated to a site in Jackson, the state capital, which is now the campus of Millsaps College.

In the early 20th century, Jackson College moved to its current location, where it evolved into a fully-fledged state university. During the Great Depression in 1934, the Baptist Society withdrew its financial support, and in 1940, the institution became publicly funded under the name Mississippi Negro Training School. It underwent several name changes to reflect its growth: Jackson College for Negro Teachers in 1944, Jackson State College in 1967 after desegregation, and Jackson State University in 1974, following the introduction of graduate programs and an expanded curriculum.

As the host institution, JSU played a pivotal role in providing a foundation for Iyengar's initiatives. This collaboration not only benefited JSU students but also set the stage for reaching out to a wider network of institutions.

- **Alcorn State University (Historically Black Institution) (Founded 1871)**

Alcorn State University holds the distinction of being the nation's first Black land-grant institution. It was founded during Mississippi's Reconstruction era by a Republican-led legislature that strongly supported the education of formerly enslaved individuals. The university was established on the grounds of the former Oakland College, which had ceased operations during the Civil War. Alcorn began its journey with three historic buildings that remain a key part of its legacy.

Iyengar's involvement with Alcorn State University exemplified his commitment to historically black institutions. Through workshops and virtual tools, he aimed to empower students and educators with the necessary skills to excel in computer science.

- **Millsaps College (Private White University) (Founded 1889)**

Millsaps College was founded in 1889–90 by Major Reuben Webster Millsaps, a Confederate veteran, who donated both the land for the college and $50,000 to establish it. The college's first president was William Belton Murrah, and Bishop Charles Betts Galloway of the Methodist Episcopal Church South led the early fundraising efforts. Both Murrah and Galloway were honored with buildings named after them. Major Millsaps and his wife are buried in a tomb located near the campus center. The college remains affiliated with the United Methodist Church today.

Collaborating with Millsaps College showcased Iyengar's dedication to bridging gaps between diverse educational settings. His work aimed to transcend barriers and offer quality education to students across backgrounds.

- **Tougaloo College (Private Historically Black Institution) (Founded 1869)**

Tougaloo College is a private, historically Black institution located in the Tougaloo area of Jackson, Mississippi. It is affiliated with both the United Church of Christ and the Christian Church (Disciples of Christ). Founded in 1869 by Christian missionaries from New York, the college was established to educate freed slaves and their descendants. Between 1871 and 1892, it operated as a teachers' training school with funding from the state of Mississippi. In 1998, the original campus buildings were added to the National Register of Historic Places. Tougaloo College is also known for its rich history of civic and social activism, including the efforts of the Tougaloo Nine.

Iyengar's efforts at Tougaloo College exemplify his commitment to historically black institutions. Through workshops and enhanced teaching approaches, he aimed to empower students who might have been traditionally marginalized.

- **Mississippi Valley State University (Founded 1950)**

Mississippi Valley State University (MVSU), also known as The Valley, is a public historically Black university located in Mississippi Valley State, near Itta Bena, Mississippi. It is a member of the Thurgood Marshall College Fund.

The university was established in 1950 as Mississippi Vocational College, following legislation passed by the Mississippi Legislature and signed into law by Governor Thomas L. Bailey on April 5, 1946. The institution officially opened on February 10, 1950, with Governor Fielding L. Wright serving as the keynote speaker at the opening ceremony.

At the time, the legislature anticipated the impending challenges to legal segregation in public education, which was later ruled unconstitutional by the U.S. Supreme Court in the 1954 Brown v. Board of Education decision. The creation of Mississippi Vocational College aimed to provide an educational alternative for African-American students who might have otherwise sought admission to the state's historically white institutions, such as the University of

Mississippi, Mississippi State University, and the University of Southern Mississippi.

By extending his reach to Mississippi Valley State University, Iyengar demonstrated his commitment to reaching students across the state, regardless of their geographical location.

- **Mississippi College (White) (Founded 1826)**

Mississippi College (MC) is a private university in Clinton, Mississippi, affiliated with the Mississippi Baptist Convention. Founded in 1826, it is the second-oldest Baptist-affiliated institution in the United States and the oldest college or university in Mississippi.

On January 24, 1826, the college received its first charter, signed by Governor David Holmes of Mississippi. Originally named Hampstead Academy, it was renamed Mississippi Academy in 1827 at the request of the board of trustees. By December 18, 1830, it became Mississippi College, offering degrees in arts, sciences, and languages.

In 1831, Mississippi College made history as the first coeducational institution in the U.S. to award degrees to female students, with Alice Robinson and Catherine Hall being among the first recipients. Although the college was not initially church-affiliated, it was connected with Methodist and Presbyterian churches for several years. Since 1850, it has been affiliated with the Mississippi Baptist Convention, with the board of trustees overseeing its operations.

Iyengar's collaboration with Mississippi College underlined his belief that quality education should transcend racial boundaries. He aimed to bring his expertise to students irrespective of their background. Dr. Paul Ohme was a Mathematics Professor and a coordinator for this program.

- **Hinds Junior College (White College) (Founded 1903)**

Hinds Community College is a public institution with its main campus located in Raymond, Mississippi, and additional branches in Jackson, Pearl, Utica, and Vicksburg.

Serving the Hinds Community College District, which covers Hinds, Claiborne, Copiah, Rankin, and Warren counties, the college enrolls over 12,000 students across its six campuses, making it the largest community college in Mississippi.

Collaborating with Hinds Junior College showcased Iyengar's dedication to supporting students in community college settings. His workshops aimed to empower students seeking education through non-traditional pathways.

- **Rust College (Founded 1866)**

Rust College, a private historically Black college located in Holly Springs, Mississippi, was founded in 1866, making it the second-oldest private college in the state. Affiliated with the United Methodist Church, it is one of ten historically Black colleges and universities (HBCUs) established before 1868 that remain in operation today.

Founded on November 24, 1866, by Northern missionaries from the Freedman's Aid Society of the Methodist Episcopal Church, Rust College is among the oldest institutions for African Americans in the U.S. In 1870, the college was chartered as Shaw University in honor of Reverend S. O. Shaw, who donated $10,000 to the institution—equivalent to roughly $240,000 in 2023.

Iyengar's involvement with Rust College exemplified his dedication to historically black institutions. Through workshops and virtual tools, he aimed to empower students and educators with the necessary skills to excel in computer science.

- **Copiah-Lincoln Community College (Founded 1915)**

Copiah–Lincoln Community College (Co–Lin) is a public community college located in Wesson, Mississippi. The Co–Lin District serves a seven-county area, which includes Adams, Copiah, Franklin, Jefferson, Lawrence, Lincoln, and Simpson counties. The college offers academic courses for the first two years of four-year degree programs, as well as career and technical programs.

Copiah–Lincoln Agricultural High School was established in the fall of 1915 through a collaboration between Copiah and Lincoln Counties, located in Wesson, at the edge of Copiah County. In the summer of 1928, Copiah–Lincoln Junior College was organized. Since its founding, the college has experienced significant growth in both size and reputation, currently enrolling over 3,200 students and possessing a physical plant valued at more than $35 million. Support for Copiah–Lincoln has expanded to include five counties: Simpson County in 1934, Franklin County in 1948, Lawrence County in 1965, Jefferson County in 1967, and Adams County in 1971.

By extending his expertise to Copiah-Lincoln Community College, Iyengar underscored his commitment to reaching students across diverse educational settings.

- **Gulf Coast Community College (Founded 1912)**

Mississippi Gulf Coast Community College (MGCCC) is a public community college located in Perkinston, Mississippi. It was originally established as Harrison County Agricultural High School in 1912.

The school began as Harrison County Agricultural High School on September 17, 1911. Four years later, the northern part of Harrison County became Stone County, and both counties continued to support the institution. On September 14, 1925, with backing from Jackson County, the school was renamed Harrison-Stone-Jackson Agricultural High School and Junior College, marking the start of its Junior College offerings. In 1942, George County joined in supporting the college, which subsequently became known as Perkinston Junior College.

Iyengar's collaboration with Gulf Coast Community College showcased his dedication to supporting students in community college settings. His workshops aimed to empower students seeking education through non-traditional pathways.

- **Belhaven College (Founded 1883)**

Belhaven University (BU) is a private evangelical Christian university located in Jackson, Mississippi. Founded in 1883, the university offers a variety of traditional majors, general studies programs, and pre-professional courses in fields such as Christian Ministry, Medicine, Dentistry, Law, and Nursing.

The college opened in 1894 as Belhaven College for Young Ladies at its current location on Peachtree Street in the historic Belhaven Neighborhood of Jackson, Mississippi. The school was initially housed in the residence of Colonel Jones S. Hamilton, a Confederate veteran who became a wealthy businessman through investments in convict-run railroads. The institution was named Belhaven in honor of Hamilton's mansion, which was named after his ancestral home in Scotland.

Belhaven University was established in 1883 through the merger of Mississippi Synodical College and McComb Female Institute, with Dr. Lewis Fitzhugh serving as its first president. J. R. Preston later acquired the school and served as its president until it was destroyed by a fire in 1910. He subsequently went on to become the superintendent of schools in Mississippi.

Iyengar's collaboration with Belhaven College aimed to provide quality computer science education to students in a private higher education setting.

Dr. Iyengar's commitment to expanding the reach of his initiatives was demonstrated through his collaborations with the University of Mississippi and other institutions. By fostering a network of excellence that transcended institutional boundaries, he enriched the educational experiences of students and empowered educators to effectively teach computer science. The impact of his efforts was felt across a diverse range of institutions, solidifying his legacy as a champion of inclusivity and educational empowerment. Through his vision and actions, Dr. Iyengar left an indelible mark on the landscape of computer science education, emphasizing the importance of education in unlocking economic prosperity and capturing the American Dream for all

Dr. Iyengar's long and distinguished record of over five decades demonstrates both depth and breadth at increasing numbers and success of underrepresented members in computer science. Iyengar's papers, state his philosophy best, "Education is the foundation for economic development and prosperity. It is the access point by which groups of all backgrounds are able to capture the American Dream." Over the past 50 years he has lived that dream, opening doors for countless others, while serving in three unique institutions.

"Separate But Not Equal: The Pursuit of Equality and Acceptance in the Deep South"

The Classroom Challenge: A Journey of Resilience and Quiet Determination

In the early 1970s, as the winds of change began to sweep through the American South, deep-rooted traditions held steadfast. Dr. S.S. Iyengar, a brown-skinned man from India, found himself in a landscape vastly different from his homeland. Arriving in Mississippi to teach at Jackson State College, an institution with a rich history of Black excellence, he was acutely aware that he was entering a complex and often unwelcoming environment, where the expectation was for either Black faculty in historically Black colleges or white educators in predominantly white institutions.

As he stepped onto the campus for the first time in the sweltering summer heat, he felt the weight of being an outsider. The cicadas buzzed incessantly, but the real noise was in the silence of cautious curiosity that greeted him. He was different—not just in appearance, but in culture, accent, and background. The small Southern town was one where familiarity and tradition ruled, and where individuals who did not fit neatly into those categories faced skepticism.

On his first day in the classroom, Dr. Iyengar stood before a small lecture hall filled with students who had mostly grown up within a hundred-mile radius. Their faces reflected a mix of curiosity and hesitation, a response he had anticipated. As he introduced himself with a thick accent and a calm demeanor, he sensed their ambivalence. Some leaned in, intrigued, while others crossed their arms defensively, unsure of how to engage with this brown-skinned professor who represented a departure from the norm.

Despite the initial apprehension, Dr. Iyengar was determined not to be deterred. He recognized that the barriers before him were not merely personal but systemic, rooted in a history that had conditioned his students to expect either a Black instructor or a white one. He approached each class with a steadfast commitment to his role, pouring his energy into meticulous preparations and leveraging his deep knowledge to bridge the divide.

The classroom remained stifled in silence for weeks. While students were polite, a certain distance lingered between them and their professor. Dr. Iyengar observed the subtleties: hesitant questions, averted gazes, and the perpetual empty seats in the front rows. It was clear that, to them, he was a puzzle—a figure they didn't quite know how to accept.

Yet, Dr. Iyengar's quiet determination began to change the dynamic. A pivotal moment came during a challenging lecture on computational theory. Recognizing that many students struggled with the material, he chose to pause and deconstruct the concept into smaller, digestible pieces. He walked around the room, establishing eye contact, and gently encouraging questions. Hesitantly at first, students began to engage.

By mid-semester, the atmosphere in the classroom shifted. The initial reluctance transformed into respect. Students who once sat in the back began moving forward, actively participating in discussions. Although a few remained aloof, the majority had come to see Dr. Iyengar not just as an instructor but as a dedicated educator who cared for their learning journey.

His acceptance didn't stem solely from his knowledge; it was his humility and willingness to listen that truly resonated. He never demanded respect but earned it through authenticity and passion for his subject matter.

One afternoon, as the semester neared its conclusion, a typically quiet student approached him after class. Shifting nervously, she finally spoke, "Professor Iyengar, thank you for being patient with us. It must have been difficult to come here and teach us. But I've learned so much from you this semester."

Dr. Iyengar felt a warmth wash over him. Though he faced many hurdles since his arrival in Mississippi, moments like this affirmed his purpose. Over the years, he became a beloved figure on campus, renowned for his brilliance and kindness, yet he always remembered those early days filled with whispers and hesitant glances.

Reflecting on his journey, Dr. Iyengar understood that the challenges he faced were not just personal but emblematic of a broader struggle for equality. He had entered Mississippi as an outsider in a classroom that traditionally reflected the racial divide, yet through resilience and quiet perseverance, he carved out a place for himself. His success became a testament to the idea that education and acceptance should transcend race and background, a belief he carried forward into his life as an educator and advocate for equality.

Silent Barriers: Navigating Stereotypes and Unspoken Exclusion in the Deep South

When Dr. S.S. Iyengar first arrived in Mississippi in the early 1970s, he had no illusions about the challenges that lay ahead. The South, though beautiful in its landscape and traditions, was a place still rooted in its past, its small towns governed by a quiet conservatism that often resisted the unfamiliar. As a professor with a name and an appearance that marked him as an outsider, Dr. Iyengar knew that his journey would not be an easy one. But of all the obstacles he had anticipated, he had not expected something as simple as finding a place to live to become such a significant challenge.

The apartment hunt began almost immediately after his arrival. Armed with local newspapers, Dr. Iyengar made his way through the classifieds, circling promising ads and calling landlords to arrange viewings. The town where he taught was small, and its housing market reflected that—a handful of modest apartments and homes were available for rent, most of them in neighborhoods where everyone seemed to know each other, and newcomers were a rare occurrence.

The first apartment he visited was in a quiet neighborhood, a tree-lined street with small, single-story homes that looked almost identical to one another. It was exactly the kind of place Dr. Iyengar had imagined himself living in—peaceful, close to campus, and, according to the ad, within his budget. When he knocked on the door, a middle-aged woman answered. She greeted him with a tight smile, but as soon as she saw him, something in her demeanor changed. It wasn't hostile, but there was a sudden stiffness, a noticeable shift in her body language that Dr. Iyengar had seen before.

"The apartment," she said, "was just rented this morning."

He glanced over her shoulder at the "For Rent" sign still hanging in the window, and though the words were polite, the message was clear. Dr. Iyengar had experienced this kind of reception before—doors that opened just enough to politely decline him, reasons that always seemed to materialize as soon as his presence was acknowledged. He knew that it wasn't personal, but it didn't make the experience any less frustrating.

Dr. Iyengar thanked the woman and left without protest. He had always been a man of quiet dignity, someone who didn't believe in making a scene. He understood that certain things couldn't be forced, and he wasn't about to create a confrontation where none would serve him. Mississippi, after all, was still a place where the past cast long shadows. He had no intention of drawing unnecessary attention to himself. So, with his characteristic calm, he turned and walked away, telling himself that there would be other opportunities.

But the pattern repeated itself. In the following weeks, Dr. Iyengar visited several more apartments, each time met with the same polite but firm rejection. Sometimes, the apartment had "just been rented"; other times, there was an excuse about needing to check with the landlord. On one occasion, a landlord offered him a tour of the apartment, all the while making it clear that the place was not actually available. Dr. Iyengar could sense the reluctance in their voices, the slight discomfort in their eyes. It was never spoken aloud, but it didn't need to be. He was not like the other tenants they were used to, and that made all the difference.

Then came the afternoon that would stick with him the most. He had seen an ad in the paper for a small apartment on the edge of town in a neighborhood known for its tight-knit community. The apartment seemed perfect—affordable, close to campus, and in a quiet area that suited Dr. Iyengar's preference for peace and solitude. He called ahead and arranged to meet the landlord.

When he arrived at the address, the first thing he noticed was the "For Rent" sign still clearly displayed in the front window. The house itself was modest but well-maintained, with a small garden in the front. As he approached the door, he felt a cautious optimism—perhaps this would be the one.

The door opened, and a man stepped out, followed closely by his wife. Both of them smiled at first, but as soon as the landlord recognized Dr. Iyengar from their phone conversation, his expression shifted ever so slightly. It was the same subtle shift that had greeted him so many times before, the kind that wasn't outright rejection but carried with it a heavy sense of reluctance.

The man cleared his throat. "I'm sorry, sir, but the apartment...well, we've had some interest, and it might already be taken. You see, we weren't expecting..."

His words trailed off, but Dr. Iyengar understood what wasn't being said. He stood there for a moment, holding the man's gaze, and then nodded.

"I see," he said softly. "Thank you for your time."

As he turned to leave, the sound of a car approaching caught his attention. A police cruiser pulled up to the curb, and two officers stepped out. Dr. Iyengar glanced back at the landlord, who was now shifting uncomfortably from foot to foot.

"There's been a call about a disturbance in the area," one of the officers said, looking at Dr. Iyengar with thinly veiled suspicion.

Dr. Iyengar knew the situation all too well. A quiet call to the police, a vague mention of something being "off" in the neighborhood, and suddenly, he was standing in front of officers who were here to investigate nothing more than his presence. The white picket fences, the manicured lawns, the close-knit community—all of it had boundaries, and he had unknowingly crossed them.

But, as always, Dr. Iyengar handled the situation with grace. He greeted the officers politely, explained that he had been visiting to inquire about the apartment, and after a few tense moments, the officers nodded and drove off, and satisfied that there was no real disturbance after all.

It wasn't the first time something like this had happened, and it wouldn't be the last. Dr. Iyengar knew that his journey in Mississippi would be filled with such moments—quiet challenges that tested his patience, resilience, and determination. But he also knew that, in time, the right opportunity would come. There was no bitterness in his heart, no anger in his voice when he recounted these stories years later. Instead, there was a quiet sense of pride, a recognition that he had weathered those moments with dignity and had never allowed them to diminish his sense of purpose.

The apartment hunt continued for several more weeks, each experience blending into the next, and each door that closed reinforcing his resolve to find a place where he could live in peace. Eventually, after much persistence, Dr. Iyengar found a small apartment on the outskirts of town, rented to him by a kind older couple who seemed unconcerned with the differences that had caused others to hesitate. It wasn't the fanciest place, but it was home, and it was all he needed.

In the years that followed, Dr. Iyengar would go on to build a life in Mississippi, one defined not by the struggles he faced but by the quiet victories he achieved through patience and perseverance. The apartment hunt was just one chapter in that story, a chapter that, like so many others, revealed the strength of his character and the unshakable belief that no matter the obstacles, there was always a way forward.

Navigating Social Barriers: Dr. S.S. Iyengar's Journey to Acceptance in a Southern Community

In the mid-1970s, Dr. S.S. Iyengar, a brown-skinned professor from India, found himself navigating the complexities of life in Mississippi, a place marked by deep-seated traditions and stark racial divides. Despite the challenges he faced as an outsider, he was determined to build connections within the local community. One evening, he received an invitation from a couple—both faculty members at Jackson State College— who were active in the African American community. They invited him to their home for dinner, hoping to share their experiences and foster a deeper understanding between their cultures.

As he approached their home in a neighborhood that bore the weight of poverty, Dr. Iyengar felt a mix of anticipation and unease. The houses, with peeling paint and cracked sidewalks, told stories of hardship and resilience. He parked his car, took a deep breath, and stepped onto the porch, where the couple greeted him with warm smiles that belied the struggles they faced.

Inside, the aroma of a home-cooked meal filled the air, and the ambiance was cozy yet tinged with an underlying tension. As they sat down to eat, Dr. Iyengar admired the culinary skills of his hosts while sensing the palpable discomfort that hung in the air. The couple, while polite and hospitable, often glanced nervously toward the window, their laughter punctuated by silences that seemed to echo their unspoken fears.

Throughout the dinner, conversations flowed, but beneath the surface, there was a heaviness that Dr. Iyengar could not ignore. He noticed how the couple would shift topics whenever the discussion veered too close to race relations or their experiences with the local white community. Their discomfort stirred a sense of empathy in him, prompting him to gently probe deeper.

"Why do you seem so uneasy?" he asked, his voice calm and inviting. "You have so much to offer, yet I can sense a hesitation."

The husband exchanged a glance with his wife before responding, his voice low and measured. "Dr. Iyengar, this is the worst state in the country for discrimination towards both blacks and browns. We live in a constant state of fear."

The words struck Dr. Iyengar with the weight of their reality. He leaned in, encouraging them to share more. "What do you mean? I've faced my own challenges here, but I want to understand your perspective."

The couple shared their stories of encounters with prejudice—how they had been subjected to ridicule and suspicion in their own neighborhood, often being followed in stores or denied service at restaurants. They spoke of the systemic inequalities that pervaded every aspect of their lives, from education to employment, and how the fear of traditional whites loomed over them like a dark cloud.

"Even though we're both faculty members, we still feel like outsiders," the wife added, her voice shaking with emotion. "You wouldn't believe the things we've had to endure, the looks we get from our neighbors, the whispers when we walk down the street."

Dr. Iyengar listened intently, understanding that their experiences resonated with his own as a person of color navigating an environment steeped in historical and social divides. He felt an unspoken kinship with them, a recognition of the common barriers they faced in a society that insisted on maintaining separations, both socially and economically.

"What can we do?" Dr. Iyengar asked, genuinely curious. "How can we change this narrative?"

The couple exchanged looks filled with a mixture of hope and resignation. "We have to keep fighting, keep teaching our children the value of education and resilience," the husband replied. "But it's exhausting. Sometimes, it feels like we're up against an unyielding wall."

As the evening progressed, the conversation shifted from fear to the small victories they had experienced. They spoke of the joy they found in their community, the strength that came from supporting one another, and the pride they felt in their cultural heritage. Dr. Iyengar felt the atmosphere brighten slightly as they shared stories of activism and solidarity, of how they sought to uplift their neighbors despite the adversity they faced.

By the time dinner came to an end, Dr. Iyengar felt a deeper connection with the couple, forged not only through shared experiences but through an understanding of the broader societal forces at play. He realized that while he was seen as an outsider in this community, he was also part of a larger narrative that encompassed struggles against discrimination, fear, and inequality.

As he left their home, he felt an overwhelming sense of gratitude. The couple had opened their lives to him, allowing him a glimpse into the harsh realities they navigated daily. In turn, he promised himself that he would be an ally, using his position in academia to advocate for equality and representation for all marginalized voices.

In that small moment of vulnerability, Dr. Iyengar understood that acceptance could only be achieved by acknowledging the pain of the past while striving together for a future defined not by the separations of race and color, but by the shared humanity that binds them all.

Grace Under Quiet Scrutiny: Dr. S.S. Iyengar's Journey Through Subtle Barriers in a Foreign Land

In the early years of his life in Mississippi, Dr. S.S. Iyengar quickly realized that even the most mundane tasks were accompanied by moments of tension and quiet scrutiny. It wasn't just the classroom or the search for housing that posed challenges—it was the everyday things, the errands that most people could complete without a second thought. For him, these experiences were colored by a palpable sense of being watched, of standing out in a place where the unfamiliar wasn't easily embraced. Yet, as always, he faced these moments with grace, his resolve unshaken by the subtle barriers he encountered.

One such memory stood out in Dr. Iyengar's mind—a simple trip to the local grocery store that had become, over time, a regular part of his routine. The store itself was a modest establishment, the kind of place where the aisles were narrow, the shelves lined with just enough variety to meet the needs of the small town's residents. It was always well-lit, with the comforting hum of refrigeration units in the background and the familiar scent of freshly baked bread greeting customers at the entrance.

It was a typical sunny afternoon when Dr. Iyengar entered the store, his list of essentials folded neatly in his pocket. As he stepped inside, he was immediately aware of the eyes that followed him. It wasn't overt or hostile, but there was always a sense of quiet observation whenever he walked into the store.

He had become used to it by now—the way customers would glance at him from the corners of their eyes, the way the staff seemed to take note of his presence the moment he entered. No one ever said anything, but he could feel it. There was an unspoken curiosity about him, about his appearance, his accent, and perhaps most of all, his place in their town.

Dr. Iyengar moved through the aisles, picking up the items he needed with his usual methodical precision. A carton of milk, a loaf of bread, some fresh vegetables—nothing out of the ordinary. Yet, as he browsed, he couldn't shake the feeling that he was being quietly monitored. It wasn't paranoia but rather the recognition of a pattern he had come to know all too well. Customers would walk by, offering polite nods or avoiding eye contact altogether. The store's employees would glance at him, then quickly turn their attention back to their tasks, though he often caught them watching him from a distance.

He reached the produce section, carefully selecting a few apples. As he inspected the fruit, he noticed the store manager standing near the back, pretending to adjust the display of canned goods while occasionally glancing in his direction. It wasn't the first time Dr. Iyengar had seen him do this. The manager, a stout man with a neatly trimmed mustache and a slightly nervous energy, always seemed to appear whenever Dr. Iyengar was in the store. He never said anything—never greeted him, never engaged him in conversation—but his presence was always felt. It was as though he was waiting for something, though Dr. Iyengar couldn't say what.

With his basket full, Dr. Iyengar made his way to the checkout counter. The cashier, a young woman with a nervous smile, greeted him with a hesitant "hello" as she began to scan his items. Her hands moved carefully, almost deliberately, as if she were afraid of making a mistake. Dr. Iyengar stood silently, watching as she double-checked each item before placing it in a paper bag. There was nothing wrong with her actions—she was simply being thorough—but there was something about the way she moved, the way she avoided his gaze suggested a deeper layer of uncertainty.

As she scanned the last item, Dr. Iyengar offered a polite smile and handed her the exact change. "Thank you," he said, his voice calm and measured, as it always was.

The cashier seemed momentarily startled by his words as if she hadn't expected him to speak. She fumbled slightly with the coins, her cheeks flushing as she hurriedly placed them in the register. Dr. Iyengar had seen this reaction before—the surprise that often accompanied the realization that he was just like any other customer, polite and unassuming. He offered another smile as she handed him his receipt, her hands trembling just enough for him to notice.

On his way out, Dr. Iyengar felt the weight of the manager's eyes on him once again. This time, the manager followed him toward the door, keeping a careful distance but making his presence known. Dr. Iyengar didn't turn around, didn't acknowledge the man's quiet surveillance. He simply walked at his usual pace, his mind focused on the path ahead. The door chimed as he stepped outside, the warm afternoon sun greeting him as he began the walk back to his apartment.

The street was quiet, lined with modest homes and a few scattered trees. Dr. Iyengar's steps were steady, his mind calm, but he couldn't help but reflect on the experience he had just had. It wasn't the first time he had felt this kind of quiet observation, and it certainly wouldn't be the last. The people in the town weren't unkind, but there was always a sense that he didn't quite belong, that he was a fixture in their community who stood out in ways they couldn't fully understand.

He had come to terms with this over the years. There was no anger in his heart, no resentment for the way people looked at him or treated him with a cautious distance. He understood that change came slowly, that the unfamiliar often made people uncomfortable, and that he represented something different in their otherwise familiar world. But he also knew that his presence in the town was important—not just for himself but for the community as well. By quietly living his life, going about his daily errands, and teaching his classes with dignity and grace, he was slowly, subtly, helping to shift perceptions.

As he neared his apartment, Dr. Iyengar allowed himself a small smile. It was a peaceful neighborhood, one that had taken some time to accept him but had eventually come to see him as a quiet, unassuming resident. The sun was beginning to set, casting a warm glow over the houses, and for a moment, Dr. Iyengar paused to take it all in. The journey to the grocery store, like so many of his other experiences, was a reminder of the quiet challenges he faced each day, but it was also a testament to his resilience.

He had come to this town as an outsider, but with each passing day, he was carving out a space for himself, bit by bit, errand by errand. It wasn't always easy, and there were moments when the weight of those curious eyes felt heavy, but Dr. Iyengar knew that he had a place here. It wasn't just about finding acceptance from others—it was about finding peace within himself, about knowing that he belonged, regardless of how others saw him.

As he unlocked the door to his apartment and stepped inside, Dr. Iyengar set the grocery bags on the counter, feeling a quiet sense of satisfaction. The journey to the grocery store, with all its unspoken tension and subtle challenges, was just one small part of his larger journey—a journey of patience, perseverance, and quiet determination. And in that journey, he had found his strength.

4.5 Life as an Academician – Pioneering Five Decades of Excellence

Dr. Iyengar's remarkable journey as an academician spans over five decades, a testament to his unwavering commitment to increasing the numbers and success of underrepresented members in the field of computer science. His philosophy, eloquently captured in his own words, underscores the transformative power of education as the foundation for economic development and prosperity. With a profound belief in the potential of individuals from all backgrounds to capture the American Dream, Dr. Iyengar has lived out this vision by opening doors for countless students throughout his distinguished career.

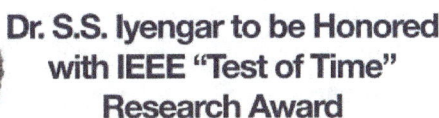

Dr. S.S. Iyengar to be Honored with IEEE "Test of Time" Research Award

"I'm deeply humbled by the praise we have received for our algorithm and astonished at the number of uses for which this technology has been applied."

As a true visionary, Dr. Iyengar's initial efforts to empower minority students took shape in the form of workshops and short courses aimed at introducing computer science to a wider audience. The Master's program in Computer Science that he initiated became a shining beacon, providing advanced education to more than 5,000 minority students. Notably, a quarter of the inaugural student cohort progressed to earn doctoral degrees from prestigious institutions such as Iowa State, University of Maryland, and Purdue. Moreover, over

60% of these graduates secured positions in Fortune 500 companies, including Lockheed Martin and Raytheon, with many holding influential roles as program directors. This landmark program also paved the way for the establishment of a PhD program in Computational and Data Science, a lasting testament to Dr. Iyengar's enduring legacy.

4.5.1 Louisiana State University (LSU) (Founded 1860)

Dr. S.S. Iyengar's Tenure - 1979-2010

In the fall of 1979, Dr. S.S. Iyengar faced a pivotal moment in his academic career as he prepared to transition from Jackson State University (JSU), a historically Black college, to Louisiana State University (LSU), a flagship institution in Baton Rouge. Despite the excitement of the opportunity, Dr. Iyengar was acutely aware of the challenges that lay ahead. Many candidates from HBCUs often faced subtle reservations in being hired by predominantly white institutions, where biases lingered regarding the caliber of education provided at Black schools.

Image: Dr. Walter Rudd, the head of the Computer Science Department at Louisiana State University, 1979

Nevertheless, Dr. Iyengar's dedication and accomplishments caught the attention of Dr. Walter Rudd, the head of the Computer Science Department at LSU, who recognized his potential and offered him a position as an assistant professor. This opportunity marked the beginning of a transformative journey, not only for Dr. Iyengar but for the institution itself.

Dr. Iyengar embraced his role at LSU with unwavering determination, setting out to elevate the university's status in the competitive landscape of higher education. Over the next three decades, he tirelessly worked to enhance the department's curriculum, foster research initiatives, and attract top-tier students and faculty. His innovative approaches and commitment to excellence played a crucial role in propelling LSU into the top 30 schools in the country by 2010. Through mentorship, collaboration, and a relentless pursuit of knowledge, Dr. Iyengar not only transformed the Computer Science Department but also laid the groundwork for future generations of scholars, ensuring that diversity and inclusion became integral to the fabric of the university's mission.

During his tenure at LSU, Dr. S.S. Iyengar remained dedicated to fostering collaboration between LSU and Southern University, a historically Black college and university (HBCU). Recognizing the need for greater diversity in academia and the importance of bridging the gap between predominantly white institutions and HBCUs, Dr. Iyengar spearheaded a groundbreaking initiative to develop a joint faculty program funded by the National Science Foundation (NSF). This innovative program aimed to promote shared resources, expertise, and research opportunities between the two institutions, creating a pathway for students and faculty from diverse backgrounds to thrive in an inclusive academic environment.

The collaborative effort proved to be a resounding success, serving as a model for similar initiatives across the nation. By fostering joint appointments and encouraging faculty exchanges, the program not only enriched the academic experiences of students at both universities but also enhanced the overall research output and quality of education offered. Dr. Iyengar's commitment to this initiative not only elevated LSU's standing as a leader in diversity and inclusion but also demonstrated how collaborative frameworks could effectively address historical inequities in higher education. The program received national recognition from the NSF, inspiring other institutions to adopt similar models and highlighting Dr. Iyengar's role as a visionary leader in the pursuit of educational equity.

Louisiana State University (officially known as Louisiana State University and Agricultural and Mechanical College, or LSU) is a public land-grant research university located in Baton Rouge, Louisiana. Established in 1860 near Pineville as the Louisiana State Seminary of Learning & Military Academy, the university's main campus was dedicated in 1926. It features over 250 buildings designed in the style of Italian Renaissance architect Andrea Palladio and spans a historic district of 650 acres on the banks of the Mississippi River.

Image: Louisiana State University, Baton Rouge, LA

As the flagship university of Louisiana and the leading institution in the Louisiana State University System, LSU enrolled more than 28,000 undergraduate students and over 4,500 graduate students across 14 schools and colleges in 2021. The university is classified as an "R1: Doctoral Universities – Very high research activity" institution and is involved in approximately 800 sponsored research projects supported by various agencies, including the National Institutes of Health, the National Science Foundation, the National Endowment for the Humanities, and the National Aeronautics and Space Administration. LSU is also one of only eight universities in the United States that offers programs in dental, law, veterinary, medical, and Master of Business Administration studies.

Dr. S.S. Iyengar's tenure at Louisiana State University (LSU) from 1979 to 2010 marked a transformative period for the university, its students, and the broader academic community. His impact, as both an educator and a leader, extended far beyond the traditional boundaries of academia. During his over two decades as Chairman of the Computer Science Department, Dr. Iyengar pioneered initiatives that not only advanced research but also laid the foundation for a more inclusive and equitable academic environment. His work fundamentally changed the landscape of computer science education at LSU, particularly for minority and underrepresented students.

Building Bridges: The Southern University and LSU Joint Research Program

One of Dr. Iyengar's most notable contributions during his tenure at LSU was the establishment of the Southern University and LSU Joint Research and Educational Program in High-Performance Networking. This groundbreaking initiative was supported by the National Science Foundation's Experimental Program to Stimulate Competitive Research (NSF/EPSCoR) and sought to provide faculty and students at historically black colleges and universities (HBCUs) with access to high-technology networks, essential for engaging in cutting-edge research.

.**Link:** https://eric.ed.gov/?id=EJ980395

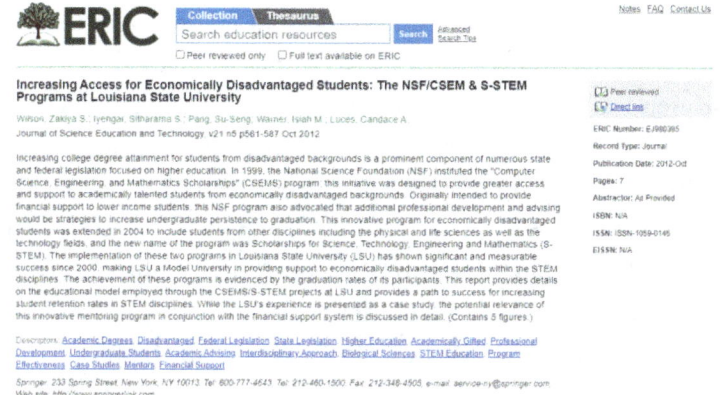

The program addressed a critical need at the time: bridging the technological divide that existed between well-funded research institutions and HBCUs. Dr. Iyengar's vision was to create a system that offered equal opportunities for all students, regardless of their background, to participate in high-level scientific research. By integrating faculty and students from HBCUs into this program, he helped open doors that had previously been closed due to a lack of resources and infrastructure.

The Southern University and LSU Joint Research Program became a model for collaboration between majority institutions and HBCUs, and it set the stage for more programs aimed at addressing the inequities in access to research and technology. The high-performance networks developed through this initiative allowed faculty at HBCUs to incorporate advanced tools and methodologies into their teaching and research, which, in turn, provided students with better opportunities to learn and apply the latest technologies.

Fostering Employment Prospects for Minority Students

Dr. Iyengar was keenly aware that the ultimate goal of these educational initiatives was not just academic success but also improving employment prospects for underrepresented students in computer science and information technology (IT). His initiatives were designed to equip students with the skills necessary to thrive in a rapidly changing technological landscape. By bridging the gap between academia and industry, Dr. Iyengar ensured that the students who participated in these programs were well-prepared for careers in high-demand fields.

The hands-on experience students gained through the high-performance networking project was invaluable. Not only did it enhance their understanding of computer science principles, but it also provided practical skills that would make them competitive in the job market. Moreover, the project encouraged collaboration among students from different backgrounds, fostering a diverse learning environment where ideas and innovations could flourish.

Mentorship and Student Success: A Legacy of Inclusion

Dr. Iyengar's impact at LSU was not limited to large-scale programs and projects; it extended to his personal interactions with students. Over his career, he supervised and mentored 50 Ph.D. students and more than 110 Master's graduates, many of whom went on to achieve success in academia, industry, and government. His mentorship went beyond academic guidance—he was deeply invested in the personal and professional growth of his students, often providing career advice long after they had graduated.

A key aspect of Dr. Iyengar's mentorship was his dedication to fostering inclusion and ensuring that underrepresented students had access to the same opportunities as their peers. His mentorship wasn't limited to his own students at LSU. Through programs like the LSU Pre-Doctoral Scholars Institute, he expanded his reach to students from other universities, particularly those from minority-serving institutions. This program was designed to prepare minority undergraduates for doctoral studies by offering intensive research experiences, mentorship, and workshops on academic writing and presentation skills.

His dedication to students was reflected in the strong bonds he formed with them, bonds that often lasted long after they had left LSU. Many of his former students credit Dr. Iyengar with shaping their careers and providing them with the tools they needed to succeed in their respective fields.

Establishing Minority Programs: Faculty Robotics Workshops and Collaborative Learning

Dr. Iyengar's efforts to advance minority education extended beyond student mentorship. Recognizing the importance of faculty development in improving educational outcomes for minority students, he played a pivotal role in the creation of faculty robotics workshops. These workshops were aimed at enhancing the capacity of minority-serving institutions to teach cutting-edge technology, particularly in the areas of robotics and high-performance computing.

By training faculty members, Dr. Iyengar ensured that the impact of his initiatives would be long-lasting. Faculty who participated in these workshops were better equipped to teach complex technological concepts to their students, thereby raising the overall level of education at their institutions. This, in turn, improved the prospects for minority students entering fields like computer science, robotics, and IT.

In collaboration with institutions such as Hampton University, the University of Puerto Rico, Southern University, Grambling University, and Morehouse College, Dr. Iyengar further extended his influence. His work with these institutions involved establishing interactive learning programs, distinguished lecture series, and collaborative projects that provided minority students with opportunities to engage in advanced research. These initiatives were critical in providing a pathway for minority students to gain the skills and experience necessary to compete in the increasingly globalized world of technology.

The Center for Research in High-Performance Computing at Morehouse College

One of Dr. Iyengar's lasting legacies was the establishment of The Center for Research in High-Performance Computing at Morehouse College. This center became a hub for faculty and students to engage in state-of-the-art research, focusing on areas critical to the future of technology, including artificial intelligence, machine learning, and computational biology.

Dr. Iyengar's collaboration with Morehouse College was more than just a research initiative; it represented a commitment to ensuring that minority students had access to the same resources and opportunities as those at larger, more well-funded institutions. By empowering faculty at Morehouse and other HBCUs to conduct high-level research, he helped elevate the academic standing of these institutions and provided a platform for minority students to showcase their talents on a national stage.

Commitment to Scholarships: Paving the Way for Future Generations

In addition to his academic contributions, Dr. Iyengar has made significant financial contributions to support the education of minority students. Every year, he donates money to establish a scholarship in his family's name, aimed at supporting students in the computer science department. This scholarship serves as a testament to Dr. Iyengar's belief in the transformative power of education.

The scholarship not only alleviates the financial burden on students but also opens doors for them to explore their full potential. By providing students with the resources they need to focus on their studies, Dr. Iyengar is helping to pave the way for future leaders in the field of computer science. His contributions to scholarships reflect his long-held belief that education should be accessible to all, regardless of their background or economic circumstances.

A Lasting Impact on Computer Science Education

Dr. S.S. Iyengar's tenure at Louisiana State University is a testament to his lifelong commitment to education, research, and the empowerment of underrepresented communities. Through his leadership, initiatives, and personal mentorship, he has transformed the lives of countless students, many of whom have gone on to make significant contributions to the fields of computer science, technology, and academia.

His work at LSU, particularly his efforts to bridge the gap between majority institutions and HBCUs, has had a profound and lasting impact on the academic community. By fostering collaborations between institutions and providing minority students with the tools and resources they need to succeed, Dr. Iyengar has created a legacy that will continue to benefit generations of students.

Under Dr. S.S. Iyengar's visionary leadership and two decades of dedicated service as Chairman, the LSU Computer Science Department witnessed tremendous growth and recognition. His relentless efforts in driving cutting-edge research, fostering inclusivity, and establishing strong collaborations with HBCUs played a pivotal role in the department's rise. Through his hard work, LSU's Computer Science Department achieved a significant milestone, being ranked among the Top 30 in the National Research Council (NRC) rankings. This achievement stands as a testament to Dr. Iyengar's unwavering commitment to academic excellence and his enduring impact on the field of computer science.

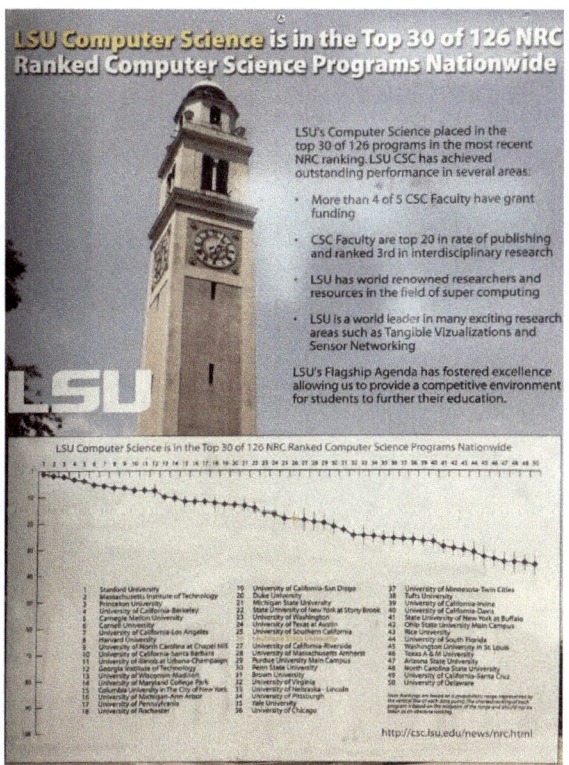

In conclusion, Dr. Iyengar's career at LSU was defined by his commitment to excellence, inclusion, and the belief that education has the power to change lives. His contributions to the university, its students, and the broader academic community have left an indelible mark, one that will continue to inspire and uplift future generations of scholars.

After dedicating 32 years to LSU, where he began as an assistant professor and rose to the esteemed position of Distinguished Professor and Chairman of the Department for 21 years, Dr. Iyengar and his wife, Manorama Iyengar, decided it was time to transition to a new opportunity at a university in Florida, specifically Florida International University (FIU), a minority-serving institution. He took on the roles of Director and Dean of the College of Computing and Information Sciences, working under the leadership of President Dr. Mark Rosenberg.

4.5.2 Florida International University (FIU) (Minority-Serving Institution) (Founded 1965)

Dr. S.S. Iyengar's Tenure - 2011-Present

Image: Florida International University, Miami, FL

FIU is Miami's public research university and a minority-serving institution (MSI). It is also the largest Hispanic Serving Institution (HSI) in the United States. Offering bachelor's, master's and doctoral degrees, both on campus and fully online. FIU is worlds ahead in its service to the academic and local community. FIU is the No. 29 university in the nation and the fourth-ranked public university, according to the America's Best Colleges 2024 rankings published on WSJ.com. Designated a Preeminent State Research University, FIU emphasizes research as a major component in the university's mission. The Robert Stempel School of Public Health and Social Work and the Knight Foundation School of Computing and Information Sciences' Discovery Lab, are just two of many colleges, schools and setting new standards through research initiatives. For more than 50 years, FIU has positioned itself as one of South Florida's anchor institutions by solving some of the greatest challenges of our time. We are dedicated to enriching the lives of the local and global community. With a student body of more than 56,000, we are among the largest universities in the nation and have collectively graduated more than 300,000 alumni, 165,000 of whom live and work in South Florida.

Dr. S.S. Iyengar's journey at Florida International University's (FIU) School of Computing and Information Sciences (SCIS) since 2011 has been marked by visionary leadership and transformative initiatives aimed at expanding access to research opportunities, fostering diversity, and empowering underrepresented groups. Over his tenure, Dr. Iyengar has continued to push the boundaries of innovation in computer science education and mentorship, significantly influencing the trajectory of FIU and its students. His efforts have not only elevated the institution but also impacted countless individuals, opening doors for African American, Hispanic, and female students, particularly in fields where they have historically been underrepresented.

Image: Dr. Iyengar with FIU President Dr. Rosenberg

**Image: Dr. Iyengar with Students at the Inauguration of
AI and Coding Club at FIU**

Fostering Research Opportunities for Underrepresented
Students

At FIU, Dr. Iyengar's focus has been on expanding access to research programs for minority and female undergraduate students, particularly through initiatives such as the **Discovery Laboratory**. This laboratory was not just a research space—it was a transformative environment where students could engage in cutting-edge projects that challenged the status quo and addressed real-world problems. Dr. Iyengar's leadership ensured that the lab offered students opportunities to participate in high-impact research, collaborate with industry partners, and develop their entrepreneurial skills.

The Discovery Lab became a launchpad for innovation and creativity, where students were encouraged to take risks, think critically, and apply their knowledge in ways that had the potential to shape their future careers. By fostering an environment that embraced interdisciplinary collaboration, Dr. Iyengar was able to guide students in combining computer science with fields like artificial intelligence, robotics, and cybersecurity. This comprehensive approach prepared students to tackle some of the most pressing challenges of the modern world, from data privacy issues to the development of intelligent systems capable of solving complex global problems.

In addition to technical training, the Discovery Lab also emphasized **student entrepreneurship**. Dr. Iyengar recognized the importance of equipping students not just with academic knowledge, but with the entrepreneurial skills needed to navigate the tech industry. Through mentorship and guidance, students developed ideas into viable business models, contributing to their long-term career growth. Many students who emerged from the Discovery Lab have gone on to found their own tech startups or hold key positions in leading companies, a testament to the powerful mentorship they received under Dr. Iyengar's direction.

Expanding Diversity and Inclusion Initiatives

Dr. Iyengar's commitment to increasing diversity in STEM fields has been one of the cornerstones of his tenure at FIU. He has long recognized the critical need for increased representation of women and minorities in technology and computer science, and his efforts have consistently sought to break down barriers that prevent these groups from thriving in these fields.

His initiatives have included **scholarship programs, mentorship schemes, and specialized outreach efforts** aimed at attracting more underrepresented students to computer science programs. At FIU, Dr. Iyengar established programs that directly addressed the challenges faced by these students, including a lack of access to resources, limited role models in the field, and systemic inequalities that disproportionately affect minority students.

By working closely with the university administration, Dr. Iyengar played an instrumental role in securing funding for initiatives that offered financial support to underrepresented students. These scholarship programs not only alleviated the financial burdens faced by many students but also ensured that those who had the passion and potential to succeed in computer science were given the opportunity to do so.

Beyond financial support, Dr. Iyengar emphasized the importance of **creating a supportive and inclusive**

environment where students from all backgrounds could thrive. Through mentoring networks, he connected students with experienced professionals and researchers who provided guidance, encouragement, and insight into navigating their academic and professional journeys. These mentorship opportunities were particularly valuable for female students and students of color, who often faced additional challenges in male-dominated fields like computer science and technology.

Pioneering Innovation at the Discovery Lab

The Discovery Lab at FIU stands as one of the signature achievements of Dr. Iyengar's tenure at the institution. This innovative space has not only prepared students for real-world challenges but also fostered interdisciplinary collaboration across a wide range of cutting-edge fields, including **artificial intelligence (AI), robotics, and cybersecurity**.

Through his leadership, the Discovery Lab became a central hub for research and innovation at FIU, providing students with opportunities to engage in hands-on projects that address critical global challenges. Students worked on AI-driven solutions to tackle issues ranging from **healthcare and climate change to cybersecurity threats and autonomous systems**. The lab fostered a dynamic environment where students were encouraged to push boundaries, think creatively, and solve problems that have far-reaching implications.

A notable example of the lab's success under Dr. Iyengar's leadership is the development of **autonomous robots and intelligent systems** designed to address real-world problems, such as disaster relief and environmental monitoring. Students at the lab have also contributed to breakthroughs in **machine learning algorithms, network security frameworks, and data privacy solutions**. These projects not only enhanced the students' technical expertise but also prepared them to become thought leaders and innovators in their respective fields.

The Discovery Lab's success was further amplified by its focus on **student entrepreneurship**. Dr. Iyengar encouraged students to take their research and innovative ideas beyond the laboratory by developing business models and launching startups. His mentorship in this area provided invaluable support to students as they navigated the complex landscape of entrepreneurship, preparing them for the challenges of starting and growing tech ventures. Several lab alumni have gone on to launch successful companies, bringing their innovations to the marketplace and contributing to the broader tech ecosystem.

Mentorship and Lifelong Commitment to Students

One of Dr. Iyengar's most enduring legacies is his deep and ongoing commitment to mentorship. His impact extends far beyond the classroom, as he has consistently demonstrated a lifelong dedication to guiding and supporting his students long after they graduate. Dr. Iyengar has made it a point to stay connected with his former students, offering guidance and advice as they advance in their careers.

His mentorship goes beyond academic advising; Dr. Iyengar is known for his **personal investment in the success of his students**, helping them navigate not only the academic challenges of graduate school but also the professional and

143

personal challenges that arise throughout their careers. His guidance has been particularly impactful for students from underrepresented groups, many of whom have faced additional hurdles in their academic and professional journeys. By providing a consistent source of support, Dr. Iyengar has helped many of his students achieve leadership positions in academia, industry, and government.

This dedication to mentorship is evident in the strong bonds he has formed with his students, many of whom credit Dr. Iyengar with playing a pivotal role in their success. His open-door policy and willingness to listen to and support his students have earned him the respect and admiration of countless individuals who have had the privilege of learning from him.

Commitment to Diversity in Technology: The AI and Coding Club at FIU

Dr. Iyengar's passion for fostering diversity and inclusion is further demonstrated through his role in the establishment of the **AI and Coding Club at FIU**, which he inaugurated as a way to **broaden access to AI education and coding skills** among underrepresented students. The club quickly became a popular initiative, attracting students from diverse backgrounds who were eager to explore the rapidly evolving fields of AI, machine learning, and coding.

The AI and Coding Club provided students with opportunities to engage in hands-on projects, attend workshops, and collaborate on real-world applications of AI technology. Dr. Iyengar's leadership and mentorship in the club encouraged students to pursue their passion for technology and develop skills that would position them for success in a competitive job market. His vision for the club was to create a space where students could learn from one another, share ideas, and work together on projects that had the potential to make a meaningful impact on society.

The AI and Coding Club has become an essential part of FIU's computer science community, providing students with the resources and support they need to succeed in a rapidly evolving field. Through this initiative, Dr. Iyengar has continued to champion **diversity in technology**, ensuring that underrepresented students have access to the tools, mentorship, and opportunities they need to thrive in fields like AI and coding.

The Impact of Dr. Iyengar's Legacy

Dr. Iyengar's contributions to Florida International University and the broader academic community have been nothing short of transformative. His **visionary leadership, dedication to diversity, and commitment to mentorship** have reshaped the landscape of computer science education at FIU, ensuring that students from all backgrounds have access to the tools and opportunities they need to succeed.

His work has had a lasting impact not only on the students he has mentored but also on the broader tech industry, where many of his former students now hold influential positions. Through his efforts, Dr. Iyengar has helped create a more inclusive and equitable environment in computer science and technology, where underrepresented students can thrive and achieve their full potential.

Conclusion: A Trailblazer in Computer Science Education

In conclusion, Dr. S.S. Iyengar's tenure at Florida International University has been marked by a tireless commitment to advancing education, fostering innovation, and promoting diversity in computer science. Through initiatives like the Discovery Lab, the AI and Coding Club, and numerous mentorship programs, he has helped transform the lives of countless students and paved the way for a more inclusive and equitable future in technology. His contributions to FIU and the broader academic community will be felt for generations to come, as the students he has mentored continue to push the boundaries of innovation and create a better world for all.

4.5 His Final Legacy: US Army Funded Forensic Investigations Network in Digital Sciences Research Center of Excellence

The increasing reliance on digital devices across society, government, and military operations has brought unprecedented challenges to cybersecurity and digital forensics. As cyberattacks and illegal penetrations become more sophisticated, so too must the field of digital forensics (DF) evolve to protect valuable information from falling into the wrong hands. The proliferation of digital devices, expected to surpass 43 billion units by 2023, adds to the complexity and urgency of securing data in an era where DF expertise is critical. In response to these growing challenges, Florida International University (FIU), a Minority Serving Institution (MSI), in collaboration with three Historically Black Colleges and Universities (HBCUs)—Grambling State University, Jackson State University, and Florida Agricultural and Mechanical University (FAMU)—has launched the **Forensics Investigations Network in Digital Sciences (FINDS) Research Center of Excellence (COE)** (https://finds.fiu.edu/)

The FINDS COE is designed to be a hub for cutting-edge research in digital forensics, specifically tailored to the needs of minority-serving institutions (MSIs). Under the leadership of Dr. S.S. Iyengar, this initiative brings together academic, government, and industry partners to collaborate on research, technology transfer, and mentoring. The goal is to build research capacity in digital forensics while enhancing the education and career development of underrepresented minority students in STEM fields, particularly digital forensics. This partnership will provide the U.S. Department of Defense (DoD), as well as industry, with advanced digital forensics capabilities and a robust pipeline of skilled professionals.

Goals and Scope of the FINDS Research Center

The FINDS Research Center has several major objectives:

1.   **Conduct Multidisciplinary, Collaborative Research Projects in Digital Forensics**
The FINDS COE is committed to pioneering research in five core areas of digital forensics, all of which focus on enhancing the ability to secure digital data and devices. These projects span a range of challenges in digital forensics, from evidence processing techniques to big data analytics. The core research themes include:

o   **Digital Forensic Analytical Methods and Evidence Processing Techniques**
Developing advanced methods for processing and analyzing digital evidence to extract actionable intelligence.

o   **Forensic Fusion Models for Extracting Event Signatures**
Utilizing network science techniques to identify and extract patterns and event signatures in digital forensics, enabling better detection and prevention of cyber threats.

o   **Analytical Methods for Big Data Digital Forensics**
Addressing the challenges posed by the massive amounts of data generated by modern digital devices, leveraging artificial intelligence (AI) and machine learning (ML) to filter and analyze data at scale.

o   **System-of-Systems Forensics for Drones and Ubiquitous Forensic Signatures**

Developing forensic methods for emerging technologies such as drones, where system-of-systems approaches are necessary to track data flows and extract forensic signatures.

o   **Workforce Development in Digital Forensics**

Creating educational infrastructure to train the next generation of DF professionals, particularly from underrepresented groups, to meet the growing demand for skilled technologists in this field.

---

2.   **Develop Enhanced Educational Infrastructure for Minority and Underrepresented Groups**

One of the key goals of the FINDS Research Center is to address the underrepresentation of minorities in the fields of computer science and digital forensics. Currently, only 8% of Hispanic and 6% of Black students in the U.S. graduate with degrees in computer science and engineering. The FINDS initiative seeks to change that by fostering diversity through innovative educational programs. By focusing on minority-serving institutions, the center provides enhanced educational tools, novel research directions, and career pathways for minority students to enter STEM careers. In particular, the center aims to increase the participation of women, Hispanic, and Black students in the digital forensics field, helping to eliminate biases in digital forensic investigations and providing more robust solutions to future challenges.

3.   Partnering with DoD, Industry, and Academia to Advance Digital Forensics Research and   Applications
The FINDS Research Center will work closely with DoD agencies, industry partners, and academic institutions to create a collaborative environment for digital forensics research and applications. By fostering technology transfer between academia and industry, the center aims to develop new forensic models, tools, and techniques that can be used by the U.S. military and other government agencies. This collaboration is essential for ensuring that cutting-edge research in digital forensics is translated into practical applications, ultimately improving operational decision-making and securing cyber systems.

Research Themes and Projects

The FINDS Research Center has developed a strategic research plan that focuses on five core projects, each of which addresses a critical aspect of digital forensics:

1.   Project 1: Towards Robust Deep Learning Systems Against Deepfakes in Digital Forensics
This project focuses on the use of deep learning systems to combat the growing threat of deepfakes, which pose significant challenges to digital forensics and security. Deepfakes involve the manipulation of

digital images, videos, and audio to create convincing but false representations. The research will develop robust AI-based methods for detecting and analyzing deepfakes, ensuring the authenticity and integrity of digital evidence.

2. **Project 2: Extracting Forensic Event Signatures Using Network Science Techniques** Network science provides a powerful framework for understanding and analyzing complex digital systems. This project seeks to apply network science techniques to digital forensics, enabling the identification of forensic event signatures from digital devices and networks. By extracting patterns and signatures, forensic investigators can better understand the sequence of events leading to a cyberattack or breach, improving their ability to prevent future incidents.

3. **Project 3: Big Data Digital Forensics** The volume of data generated by modern digital devices presents significant challenges for forensic investigators. This project focuses on developing advanced big data analytics techniques for digital forensics, using AI and machine learning to process and analyze massive datasets efficiently. The goal is to create scalable tools that can handle the growing complexity of digital forensics investigations in the era of big data.

4. **Project 4: Drone Forensics with Machine Learning-based Fingerprinting and Blockchain** As drones become more prevalent in both civilian and military contexts, they present new challenges for

digital forensics. This project will develop forensic methods for drones, utilizing machine learning-based fingerprinting techniques to track and analyze drone activity. In addition, blockchain technology will be explored as a means of ensuring the integrity and traceability of drone-related forensic data.

5.   **Project 5: Workforce Development in the Post-Pandemic   New   Normal**
The COVID-19 pandemic has fundamentally altered the landscape of education and workforce development. This project will address the challenges posed by the "new normal" by developing digital forensics training programs that are accessible to students and professionals in a post-pandemic world. The focus will be on creating flexible, online learning opportunities that provide the necessary skills for a career in digital forensics.

Building the Future of Digital Forensics

The **Forensics Investigations Network in Digital Sciences (FINDS) Research Center of Excellence** represents a bold step forward in advancing digital forensics research, education, and workforce development. By fostering collaboration between minority-serving institutions, government agencies, and industry, the center is poised to make a significant impact on both national security and the diversity of the digital forensics workforce. The research conducted at the FINDS center will lead to the development of new forensic tools and techniques, enhancing the ability of forensic investigators to secure digital devices and systems.

As the Principal Investigator of the FINDS Research Center, **Dr. S.S. Iyengar** plays a pivotal role in shaping the future of digital forensics. Through his leadership and vision, the center will provide the U.S. military, government, and industry with the expertise and tools needed to address the ever-evolving challenges of cybersecurity. At the same time, it will empower underrepresented groups to pursue careers in this critical field, ensuring that the next generation of digital forensic professionals is both skilled and diverse.

In conclusion, the FINDS Research Center embodies Dr. Iyengar's enduring legacy of innovation, collaboration, and inclusivity. His work not only advances the field of digital forensics but also contributes to a more equitable and secure future for all.

4.6 Summary: A Legacy of Transformation and Empowerment

Dr. Iyengar's profound impact on education and diversity stretches across multiple institutions, leaving an indelible mark on the academic landscape. His steadfast dedication to fostering inclusivity, research excellence, and transformative mentorship has fueled his journey of over five decades.

Image: Dr. Iyengar Mentoring Minority Students

While at Jackson State University (JSU), Dr. Iyengar played an instrumental role in the successful development and sustenance of the Master's program in Computer Science. This program, marked by its unique and innovative aspects, gained national recognition and was highlighted in the ACM SIGCSE proceedings of 1976. Graduating over 400 students, the program's graduates have achieved leadership positions in prominent technological companies and defense organizations. Dr. Iyengar's visionary leadership ensured that this program became a symbol of quality education and success for its graduates, a lasting tribute to his legacy. Dr. Iyengar's tenure at JSU also reflected his commitment to opening avenues for underrepresented students. He understood that education extended beyond traditional classroom settings, which led him to initiate workshops, short courses, and collaborations that introduced computer science to diverse audiences. His far-reaching vision was the driving force behind the expansion of opportunities for minority students, ultimately shaping their academic and professional trajectories.

154

Throughout his career, Dr. Iyengar's guiding philosophy revolved around the notion that education is not merely a means to acquire knowledge, but a transformative force that paves the way for economic empowerment and prosperity. His unwavering belief in this philosophy was reflected in his numerous initiatives, which aimed to bridge gaps, dismantle barriers, and empower individuals to realize their full potential.

Dr. Iyengar at FIU President's House

His contributions were not confined to academia alone. Dr. Iyengar's innovative projects, such as the development of high-performance networking programs, demonstrated his commitment to practical applications of technology to uplift minority communities. By providing historically underserved institutions with the tools and resources to engage in cutting-edge research and development, he facilitated greater inclusion and diversity in the tech industry.

Image: Dr. Iyengar with Mr. Adam Rogers CEO of Ultimate Software, Provost Dr. Ken Furton and College of Engineering and Computing Dean Ranu Jung

Dr. Iyengar's mentorship was not limited to the classroom; it extended to his ongoing support and guidance for his students' careers. His mentorship went beyond graduation, as he actively tracked his students' progress and offered insights and advice throughout their lives. His personal investment in their success exemplified his dedication to nurturing future leaders in computer science.

Image: Dr. Iyengar with President Dr. Rosenberg at the inauguration of FIU Tech Station

The culmination of his journey brought him to Florida International University's School of Computing and Information Sciences (FIU SCIS), where his impact continued to thrive. By increasing access to research programs and creating spaces for innovation, he ensured that underrepresented students had the tools and opportunities needed to excel in their chosen fields.

Dr. Iyengar's profound and lasting contributions across institutions, disciplines, and generations reflect his relentless pursuit of equality, education, and empowerment. His transformative mentorship, visionary leadership, and innovative initiatives have not only transformed the lives of individuals but have also reshaped the landscape of computer science education, making it more inclusive and accessible to all. Dr. Iyengar's enduring legacy is a beacon of hope and inspiration for future generations, underscoring the immense power of education to drive positive change in society.

Lessons Learned

"There are only two mistakes one can make along the road to truth; not going all the way, and not starting."
-Buddha

The vignettes demonstrate the complex racial and cultural barriers of the deep South in the 1970s as Dr. Iyengar embarked on his early career. Only the members of each generation know the trials and tribulations commonplace in their eras—what they have endured, things too painful to remember and yet too important to forget. Despite the adversity there are those who have a vision, set out for success on the road to truth, going all the way. Dr. S.S. Iyengar is one of those visionaries who despite the challenges carries the torch of education for all.

In looking back at the historically black colleges and universities/minority institutions (HBCU/MIs) it is important to remember the context of their initial charters as institutions of teacher and vocational training. Change is inherently slow, even more so when there is no clear connection to advancement and implementation of technologies that may appear to be novelties, well outside of an institution's charter. This was computer science in the 1970s.

Such was the challenge that Dr. Iyengar, a junior faculty member, faced in the 1970s with university

---

administrators, as he attempted to institute an advanced technology that only a decade earlier was not even an independent scientific discipline. Dr. Iyengar began an educational and technological journey that required his utmost efforts to implement. To do so required that he take Dr. Jesse Lewis' vision, make it his own, while developing and implementing an action plan. He continuously evolved the vision along the way.

Key but unanticipated elements became central to Dr. Iyengar's fight for implementation. These included *Facilitating Policy and Infrastructure Discussions* who are caught by surprise with the rapid development of computer science technology. In implementing the education program, he had to *Develop Comprehensive Course Guidelines* not only for a single institution but for multiple institutions across the state. His initial charter of *Training Faculty and Students* in the new technology enabled him to empower educators and learners with the necessary skills needed to leverage this new technology. Throughout his ride on the circuit, *he* provided the program ***Leadership*** in Curriculum Development, Sharing Experiences and Insights and Assessing Instructional Computing.

As a result, Dr. Iyengar learned to navigate the intricacies of university administration, as well as build a foundational connection between many universities in Mississippi that would become the backbone of Mississippi's intranet.

Chapter 5:
Life Lessons and Experiences

"I will prepare and someday my chance will come."

Abraham Lincoln

5.0 Overcoming Adversity: The Early Struggles and Foundations of Resilience

Dr. S.S. Iyengar's early life was marked by significant challenges, as he grew up in a period where poverty was a constant reality. Despite the hardships, he demonstrated an exceptional level of perseverance, determination, and focus that would later define his illustrious career. His family, like many others at the time, faced financial difficulties, and he took on the responsibility of contributing to their well-being. Yet, despite these circumstances, education remained at the forefront of his aspirations. In those formative years, he walked 15 kilometers every day to attend school at National College, a testament to his unwavering commitment to self-improvement and academic excellence.

Such a journey required not just physical endurance but mental fortitude as well. Walking these distances daily through difficult terrains, and in a country where infrastructure was still developing, wasn't an easy feat. However, Dr. Iyengar found a sense of purpose and strength in these struggles. His resolve to overcome obstacles was deeply rooted in the belief that education would be the key to uplifting not only himself but also his family. This dedication to knowledge is something that would become a cornerstone of his philosophy as an academic and mentor later in life.

Growing up in a family that valued traditional principles also shaped his worldview. His early years were spent not only on academic pursuits but also on developing a moral compass grounded in integrity, honesty, and respect for others. Even as a young man, he adhered strictly to these principles, which would guide his decision-making throughout his career. These values were non-negotiable for him, and they stood as the foundation upon which he built his professional and personal life. His ability to remain true to these values, even in the face of adversity, is a reflection of his inner strength.

5.1 Standing by Principles

Throughout his career, Dr. Iyengar's steadfast commitment to principles was more than a moral stance; it was a guiding force that shaped his interactions, decisions, and problem-solving approaches. He believed in resolving issues with a clear sense of right and wrong, never compromising on his ethics. Whether facing professional challenges or personal difficulties, he would often return to the idea that integrity must never be sacrificed for convenience. This mindset is one of the reasons why he was able to earn the respect and admiration of colleagues, students, and collaborators throughout his career.

Dr. Iyengar's principles were not just theoretical constructs; they were actively put into practice in his daily life. For instance, he believed in fairness and equity, always ensuring that opportunities were made available to those who were deserving, irrespective of their background. This belief in the power of equal opportunity was especially evident in his advocacy for underrepresented groups in the field of technology. His strong principles also extended to the way he approached teaching and mentoring. He believed that as a professor, it was his duty to inspire and nurture the next generation of thinkers, not just impart technical knowledge. This holistic approach made him not only a great academic but also a role model for his students.

His ability to stand by his principles often required immense personal sacrifice. In his early career, there were moments when adhering to his ethical beliefs meant turning down lucrative offers or resisting the temptation to take easier paths. He believed that success achieved without integrity was hollow, a belief that became one of the pillars of his career. It was this integrity that made him a trusted figure in both academia and industry. Even as he rose to positions of prominence, he never lost sight of the values that had guided him in his early days. His decisions were consistently informed by his commitment to fairness, honesty, and the greater good.

5.2 Career Trajectory: Twists of Fate and Good Fortune

As with many great lives, Dr. Iyengar's path was not without its twists of fate. Despite his talents and work ethic, his career wasn't a straight line to success. He faced numerous challenges, from navigating the academic world to overcoming societal and financial barriers. However, he never viewed these obstacles as deterrents; rather, he saw them as opportunities to grow and learn. This perspective is a hallmark of his personality—an ability to turn challenges into stepping stones for greater achievements.

In his view, one must not simply wait for opportunities but instead create them. Dr. Iyengar's career is a reflection of this proactive approach. Throughout his journey, he was strategic in identifying and seizing opportunities that aligned with his long-term goals. At times, this meant taking risks and making unconventional choices. For example, early in his academic career, he chose to delve into areas of research that were not yet mainstream, believing that they held significant potential for the future. His ability to foresee the relevance of emerging technologies like artificial intelligence and cybersecurity has been a key factor in his success.

Luck, as Dr. Iyengar himself acknowledges, played a role in his trajectory. However, his definition of luck was not merely random chance. He believed that luck favors those who are prepared and willing to take advantage of the opportunities presented to them. His career is filled with examples of such moments—when a fortuitous connection, a timely breakthrough, or an unexpected challenge propelled him to new heights. Yet, none of these moments would have been possible without his readiness to act and his meticulous planning.

One particularly pivotal moment came when he made the decision to collaborate internationally, long before globalization became the norm in academia. This decision opened up a wealth of opportunities for cross-border research and collaboration, further enhancing his reputation as a thought leader. His willingness to embrace new challenges, coupled with his strategic approach, allowed him to navigate complex academic and professional landscapes with relative ease.

5.3 A Game Plan for Excellence

Dr. Iyengar's philosophy on success is encapsulated in his belief that one must have a game plan for reaching the pinnacle of excellence. He emphasizes that success is not a matter of luck alone but the result of careful planning, determination, and continuous self-improvement. Throughout his career, he has consistently applied this principle, whether in research, teaching, or leadership roles. His approach to problem-solving is methodical, focusing on long-term objectives rather than short-term gains.

This mindset is particularly evident in his contributions to academia. Dr. Iyengar was never satisfied with merely achieving personal success; he aimed to build institutions and programs that would have a lasting impact. Whether it was through mentoring students, developing cutting-edge research initiatives, or advocating for diversity and inclusion in STEM fields, his focus was always on creating a legacy that would outlast his individual achievements.

His game plan for excellence also involved a commitment to lifelong learning. Even after decades of experience, Dr. Iyengar remained a student at heart, always eager to learn from new developments and adapt to changing circumstances. This humility and willingness to grow has been one of the secrets to his sustained success.

Dr. S.S. Iyengar's life is a testament to the power of perseverance, principles, and a well-thought-out game plan. His journey from humble beginnings to becoming a world-renowned academic and leader in technology showcases not only his personal resilience but also his unwavering commitment to excellence. The life lessons he has imparted, both through his actions and teachings, continue to inspire those who follow in his footsteps.

5.4 The Path of Resilience and Virtue

Dr. Iyengar's journey was one defined by perseverance, principled living, and unwavering commitment to his beliefs. Despite facing immense challenges and hardships, he embarked on a path of learning and earning through the crucible of adversity. In his formative years, Dr. Iyengar had to navigate a world where resources were scarce, and opportunities were limited. Hunger may have been his companion, but his principles and virtues emerged as the guiding lights of his life. These principles became the foundation upon which he built his character and forged his identity. Anxiety and fear were no strangers to Iyengar's life. However, instead of succumbing to their weight, he harnessed these emotions as catalysts for growth. He learned that honesty and being helpful to others not only carried intrinsic value but also yielded rich dividends in his career. Through his trials, he discovered that the path of integrity and benevolence was a true source of strength.

Image: Dr. Iyengar with Wife Manorama and Son Veneeth

Iyengar's compass always pointed towards assessing the goodness in people and wishing them success in their endeavors. His largeheartedness extended beyond himself, as he consistently demonstrated a willingness to go out of his way to help others. He intimately understood the meaning of hardships, and this awareness fueled his empathy and compassion towards those facing their own challenges. The recognition bestowed upon him through awards served as a motivating force, propelling him to continually improve in his career. His maxim, "Match your talents to your ambition," encapsulated his pursuit of growth and excellence. Each accolade became a stepping stone in his journey, reinforcing his dedication to his craft.

As he observes the younger generation's confidence, he also acknowledges a concern for the lack of resilience and compassion in today's world. While he applauds the expressive nature of contemporary society, he laments its short-sightedness. Dr. Iyengar's wisdom underscores the importance of balancing confidence with the ability to persevere through challenges and show empathy to others. Living with hope and passion, he discovered the beauty of leading a life of simplicity. Dr. Iyengar's philosophy teaches that in embracing these virtues, one can find fulfillment amidst life's complexities.

Image: Dr. Iyengar with Wife Manorama

Family values have always held a critical place in his life. He recognizes that they form the bedrock of prosperity and success. In his own journey, family values provided the essential support and grounding that allowed him to rise above adversity and become an influential figure in academia. Dr. Iyengar's story embodies the embodiment of resilience, integrity, and compassion. His life's lessons illuminate the power of virtuous living, unwavering dedication, and a heart that extends to others. Through the prism of his experiences, we glimpse the transformative potential of fortitude and principled existence.

5.5 The Essence of Diligence: Promises and Perils

The cornerstone of a successful journey lies in cultivating discipline from the outset, leading by example and shaping one's life with purposeful intention. While short-term pleasures may offer fleeting happiness, they often conceal long-term traps and disadvantages. Succumbing to immediate gratification can divert one from the path of sustained growth, undermining the potential for a fulfilling future. In contrast, embracing the challenges and enduring the rigors of hard work can pave the way for lasting success.

The comfort of an easy path can lull individuals into complacency, rendering them devoid of the motivation to strive for advancement in their careers. Complacency breeds stagnation, while the pursuit of excellence demands an unrelenting commitment to constant improvement. The mind is a tool, a reservoir of analytical skills and artistic talents waiting to be honed. A disciplined work ethic is the key to unlocking these abilities. Wasting time squanders the opportunity to cultivate one's potential, limiting the array of choices available. The onus lies on individuals to utilize their time wisely, harnessing their abilities to navigate an array of opportunities.

Opportunities are not exclusive; they are open to all. However, the crucial factor is seizing them at the opportune moment. Procrastination or hesitation can lead to the loss of valuable prospects, underscoring the importance of timely and decisive action. In essence, the belief of Iyengar that *"nothing comes easily in life, and if it does, it might not be worth it"* encapsulates the wisdom of acknowledging the value of hard work and the satisfaction derived from overcoming challenges. The promise of a fulfilling journey hinges on the willingness to put in the effort, persist through hardships, and embrace the transformative power of diligence.

5.6 The Harmony of Teaching and Research Iyengar's Philosophical Insights

- **Cultivating Scientific Thinking:** Iyengar's philosophy underscores the significance of intertwining teaching with foundational scientific exploration. He believes that researchers should actively engage students in the realm of foundational science, guiding them to embrace the same scientific thinking that fuels the discovery and comprehension of knowledge. This approach not only imparts theoretical understanding but also empowers students with the cognitive tools essential for robust intellectual exploration.

Inquiry-Based Learning for Minority Students: When it comes to minority students, Iyengar champions an inquiry-based research pedagogy. He recognizes that such an approach not only fosters a deeper understanding of subject matter but also nurtures critical thinking and analytical skills. By encouraging students to question, investigate, and discover, this method empowers them to navigate the complexities of academia and prepares them for the intricacies of research.

- **Collaboration and Interdisciplinarity:** Iyengar underscores the power of team-based discussions as a cornerstone of education. Collaborative approaches cultivate a sense of unity and camaraderie, facilitating interdisciplinary work. In an age where complex challenges require multidimensional solutions, fostering collaboration among students from diverse disciplines enriches their learning experience and equips them to tackle real-world complexities.

Final Quotes that Illuminate the Path

- *"Risk is the price you pay for opportunity."* The pursuit of progress inherently involves venturing into the unknown, taking calculated risks to seize opportunities that lie beyond familiar territory.

- *"Unless you treat failure as part of the journey, you are never going anywhere."* Embracing failure as an intrinsic part of the learning process is pivotal. Failures not only provide valuable lessons but also contribute to personal and professional growth.

- *"Don't let ideology undermine the growth of your career."* Ideological biases can potentially hinder growth by limiting one's perspectives and openness to new experiences. Remaining open-minded and adaptable is key to thriving.

- *"As Lincoln said - 'Don't worry when you are not recognized but strive to be worthy of recognition'."* Recognition is a byproduct of sustained effort and excellence. Focusing on being valuable rather than seeking validation ensures a genuine and enduring impact.

Dr. Iyengar's philosophy embodies the essence of holistic education, uniting the realms of teaching and research to nurture a generation of curious, analytical, collaborative, and resilient individuals poised to contribute to the ever-evolving world of knowledge.

Dr. S.S. Iyengar's journey began at Jackson State University (JSU), where he was offered his first job as a circuit rider in 1974. Over the next five decades, he built a distinguished career across multiple institutions, including JSU, Louisiana State University (LSU), and Florida International University (FIU), while leaving a global impact on the fields of computer science and education. His work has spanned from creating educational networks at JSU to elevating LSU's computer science department to national prominence, and finally, to leading groundbreaking research initiatives at FIU. Now, in a full-circle moment, Dr. Iyengar is returning to the institutions that shaped the early years of his career, JSU and Grambling State University, through his FINDS Grant. This final journey is a testament to his enduring commitment to giving back to the Historically Black Colleges and Universities (HBCUs) that first gave him the opportunity to make a difference. By doing so, Dr. Iyengar continues to empower the communities that played a pivotal role in his extraordinary professional life, ensuring that future generations benefit from the legacy he has built.

Lessons Learned

"If I have seen further it is by standing on the shoulders of giants."
-Sir Isaac Newton

"Civilization owes its greatest debt not to visionaries but to those who take the vision, make it their own, and invest sweat equity to make it a greater reality than envisioned."
-Jerry F. Miller

Uncompromising principles, perseverance, resilience, and vision have been the hallmarks that defined Dr. S.S. Iyengar since his early days in India. On his journey, he has become a Distinguished University Professor, a Distinguished Chaired Professor, and extraordinary researcher, and academician. He is an IEEE Life Fellow, ACM Fellow, AAAS Fellow, NAI Fellow, AIMBE Fellow, SDPS Fellow, AAIA Fellow, and member of both the European Academy of Sciences and the European Academy of Arts and Sciences. His list of other honors and awards is long and distinguished.

Throughout his career he has pushed the boundaries of knowledge. A visionary, Dr. Iyengar never hesitates to develop and implement his vision while providing a pathway for others to follow.

While he would be the first to admit that like Sir Isaac Newton, he has seen further by standing on the shoulders of giants. Through perseverance, hard work and

172

dedication, Dr. Iyengar has now become a giant himself in computer science. More importantly, he has become a giant in the field of education for minority and underrepresented students. As a circuit rider he laid the foundation for computer science education in Mississippi.

On the grounds of the United States Air Force Academy sits a statue known as the "Eagle and Fledglings." Inscribed on the monument is this quote by Austin Miller, "Man's flight through life is sustained by the power of his knowledge." It is a fitting tribute to those who constantly seek knowledge then willingly mentor others.

Without a doubt Dr. S.S. Iyengar's flight through life has been sustained by his insatiable quest for knowledge. Along the way he has fueled the flights of innumerable students, faculty, and colleagues in their quests for knowledge and will continue to do so for many more years.

Chapter 6:
A Legacy of Resilience and Vision

"Through resilience and vision, he turned adversity into opportunity, from his humble beginnings in India to pioneering computer science education in America. His journey as a 'circuit rider' for underserved communities forged a path for equitable access to technology. His legacy inspires future generations to embrace challenges as gateways to growth and positive change."

--Dr. R.L. Kashyap

"A pessimist sees the challenge in every opportunity and an optimist sees the opportunity in every challenge". My grandfather, Dr. SS Iyengar is a living example and embodiment of this ethos. Where others only saw insurmountable odds that stopped and halted their progress, he used each roadblock as not only a hurdle to overcome but also a learning opportunity in life. He not only recognized his own inherent strengths at an early age and optimized his potential but continued and continues to push others to adopt this philosophy in life.

When he was born in a small mud hut on the banks of the river Cauvery in Southern India, barely 10 days after the country gained independence (1947) from British rule, no one in his family could have foreseen the extraordinary life he would lead. Growing up in poverty, his family survived on his father's modest government salary. Yet, his strong work ethic and relentless optimism helped him excel academically, making him one of the top students at a premier engineering college in India. He went on to earn his master's degree from the prestigious Indian Institute of Science (IISC) in Bengaluru.

Though he could have stayed in India and led a comfortable life, he chose to move to the United States to pursue advanced studies in Engineering. This was during a period when very few Indian immigrants were venturing to the U.S. for education and career opportunities.

Arriving in America as an immigrant with limited resources, he found himself in the deeply segregated southern state of Mississippi during the post-Jim Crow era. In my school (Dallas and New Jersey), I learned that the Jim Crow era in the United States spanned from the end of Reconstruction in 1877 through the civil rights movement of the 1950s. This period was marked by hardship and discrimination against non-white populations. He faced financial hardships and cultural challenges, earning his Ph.D. in 1973 while balancing multiple jobs to support himself, including roles at Zenith, as a research student at MSU, and other companies. His professional journey began in January 1974 as a Circuit Rider at Jackson State University (HBCU), where he was responsible for teaching programming to students at colleges throughout Mississippi. This marked the beginning of his experience in understanding and collaborating with non-white individuals. Eventually, he obtained a faculty position at Louisiana State University (LSU), where he devoted nearly thirty years to the institution, including almost twenty years as chair of the Computer Science department.

His passions were education and research. He was among the wave of pioneers in the field of computational science, who were prescient enough to foresee how ubiquitous computers would become.

Computing devices and the internet are among the greatest inventions of humankind. No other invention has been comparable in how it has revolutionized and revamped every aspect of human life. The early days of Computer science were characterized by collaboration across mathematics, engineering and logic. All of these formed the basis for the digital revolution that transformed the 20th and 21st centuries. As more and more universities in the US started recognizing computer science as a separate and vital discipline, men and women like my grandfather took the leap into this a yet-unknown and undeveloped field of study. Researchers started exploring areas like algorithms, data structures and even artificial intelligence, setting the stage for the rapid growth of computer technology. It is thanks to all of them, their hard work, ingenuity and intellectual curiosity that humans now live in the best period of the

history of humankind. Technological advances have taken us farther at a faster pace than ever before.

As you know, the title of this book is Circuit Rider. I admit, before I started researching this topic, I really did not know anything about the origins of circuit riders. Circuit riders were clergymen, typically Baptists or Methodists who traveled from church to church within an area of service known as a circuit.

They were preachers who traveled vast distances on horseback to spread their message in underserved areas and communities in the 18th and 19th centuries. Over time, this practice was adopted by many professions, from judges to medical workers to various technical experts, as a means of meeting the needs of communities that did not have full-time professionals serving them.

My grandfather worked as a circuit rider in Mississippi. Just imagine.....a recent immigrant who spoke fluent English but with a heavy Indian accent, a non-Caucasian driving the backroads of a still segregated southern state. He faced his many apprehensions and fears head-on. He was no stranger to the concept of discrimination in his native India. While he was from an economically disadvantaged family, he was a member of the higher-caste Brahmin community.

He was never discriminated against in India but he disdained the very concept when he saw how some of his classmates and friends were treated. In Mississippi in the early 1970s, he was often the recipient of subtle reminders that he was a person of color. However, the job was an opportunity to not only support himself financially but also help universities in the area with computer maintenance, teaching their staff computer literacy etc. He worked with many HBCUs during this time. He was living proof that education was the main pathway for those on the lower rung of the socio-economic scale to rise up and build a better life. He wanted to equip the underfunded HBCUs and their students and faculty with the tools they would need to build successful computer science departments.

I was born in 2009 in Houston, Texas, one of the youngest members of GenZ. Those of us born after 2007 or so have never known a life without electronic devices at our fingertips. The laptop boom had taken hold; Blackberry phones had been knocked off their perch by the stylish and easy-to-use iPhone. We never lived through the slow days of dial-up internet connectivity. In the developed world, internet connections in homes and businesses are starting to be treated like other basic utilities such as electricity, gas, or water. From the time I was old enough to form memories, I saw large and small screens everywhere.

When my mom tells me about her childhood and how only a lucky family or two in their neighborhood would own a basic television set, it's like listening to stories about life on Mars. When I hear about how my dad (currently a famous Doctor in a very prestigious hospital in America) and his brothers would travel 30 hours from Baton Rouge to India without movies or video games to distract them, I am amazed! For good or bad, my generation and those younger than me have never known and will never know what it was like to not have information, commerce or entertainment at our fingertips!

One of the most transformative features of computers and the internet is their ability to facilitate connectivity and communication, essential elements for us as social beings. Some of our most treasured memories are made in the company of those we love, and technology has increasingly enabled us to stay close, regardless of physical distance.

When my grandfather first moved to the United States in the late 1960s, reaching his family in India meant writing letters that would take over two weeks to arrive—and that's if they didn't get lost or misdelivered along the way. At that time, international calling was prohibitively expensive and nearly impossible for families in India to access. The wait to hear back from his family required the same long and unreliable process in reverse.

As my grandfather studied and worked in the emerging field of computer science, computers were still in their infancy—large, power-hungry machines with no capacity for the types of communication we take for granted today, like instant messaging or email. Little could he have imagined that one day these early machines would evolve into tools that could instantly connect people around the world, bridging distances in a way that felt nearly impossible during his early years.

However, he did see the potential of computers as a means of communication and their ability to bring people and communities together. Now, more than 50 years after he started working as a circuit rider, generations are reaping the rewards of all the ideas that were conceived by him and other computer scientists of his generation.

Once again, for someone like me, it's impossible to imagine a world where I cannot reach my loved ones instantly just by tapping on a keyboard. Expedient communication has made every aspect of our lives easier. Families and friends who are spread across the globe can keep in touch very regularly. Work communication methods, from emails to video calls, have made all workplaces more efficient and allowed employees to have a better work-life balance.

Telehealth visits with medical professionals have improved healthcare access for all patients, especially those living in remote or underserved communities. These patients are now able to meet with a physician or nurse virtually instead of having to travel long distances. Without all these hardware and software systems in place, from personal computers, mobile phones and laptops to emails, Facetime, zoom, etc, the world would have been forced to completely shut down when Covid hit in the spring of 2020.

Computers and the internet became essential tools for adapting to a new way of life, enabling unprecedented education, healthcare, work, and social interactions.

The internet, mobile devices and social media have also been drivers of social and political justice. Ordinary people on the ground are able to get news and information out to the world through these mediums. This has been something especially vital when autocratic governments have shut down more traditional media like newspapers and TV channels. Activists and even regular citizens have been able to communicate, organize and broadcast their struggles to a global audience. From the Arab Spring in 2010 to the most recent wars in Ukraine Israel and Gaza, social media has made distant struggles seem more relevant to all our lives.

We now live in the Information Age. Information and knowledge have become the central resource, driving economies, innovation and social change. The Internet became widely available to the public by the mid to late 1990s and led to the explosion of ecommerce, social networks, informational websites and instant communication. The internet era has led to equity in information access. What this means is that with reliable and affordable access to technology, every individual, irrespective of their geographical location or socioeconomic status, has access to relevant information. Problems with misinformation persist and teaching young people to differentiate between facts and lies is more important than ever. This information access is vital in breaking down the barriers of disparity in education, economic opportunities, healthcare access etc.

My grandfather was an economically disadvantaged student who had to fight for every resource and opportunity. His work as a circuit rider aimed to equitably distribute the technological resources available at the time. His work enabled poorer universities that served communities of color to have similar resources as other places of learning.

Every person has role models and inspirational figures in their life. Admiration and appreciation of an individual's work or the way they lead their life is part of human nature. However, it is impossible to replicate someone's life experiences or achievements. My grandfather has been a part of my life since the day I was born. I have watched him and observed his passion for his work and his family, his optimism and can-do attitude, his work ethic and his desire to serve society with his skills and talents. W.E.B DuBois said, "Children learn more from what you are, not what you teach". He has shown not just me but everyone who has ever come in contact with him what it means to be a positive force in the world. As I look towards my future and what I want to achieve as I move through life, he remains one of my role models and greatest influences. His impact on me will be lasting, empowering me to overcome challenges, strive for excellence and contribute positively to society.

Epilogue: A Circuit Rider's Legacy

In December of 1973, a young man stepped off a bus in Mississippi, carrying with him little more than a suitcase and a vision—a vision to learn, educate, inspire, and uplift communities that had long been forgotten. That man was Dr. S.S. Iyengar, and he would go on to shape the future of minority education in ways no one could have imagined. Known to many as a "circuit rider"—a title given to those who travel to bring knowledge and hope to those in need—he took up the mission of empowering students in historically Black colleges and universities (HBCUs) across the Deep South. But Dr. Iyengar was no ordinary educator. His journey was not just about teaching; it was about changing lives, breaking barriers, and sowing the seeds of opportunity where others saw only obstacles.

Fifty years later, in September of 2024, I (Yashas Hariprasad) had the honor of retracing his footsteps. Together, Dr. Iyengar and I visited the same places where his incredible career first took root—Jackson State University (JSU), Grambling State University (GSU), Mississippi State University (MSU) and other places. What began as a simple tour of these institutions soon turned into something far deeper. As I walked beside him through the campuses, I could feel the weight of history pressing against us. This was not just a man revisiting the past; this was a pioneer standing on the foundation of a legacy-built brick by brick over five decades.

Jackson State College is where it all began for Dr. Iyengar. In 1973, fresh out of his PhD program at Mississippi State University, he accepted a job at JSU, earning a humble $11,000 per year. It wasn't just a paycheck; it was the beginning of a calling. He traveled across Mississippi, teaching programming to minority colleges with scarce resources and even fewer opportunities. For him, JSU was more than just an institution—it was a gateway to a lifetime of service. He described those early days as some of the most challenging but also the most rewarding of his life. He wasn't just teaching students; he was giving them a path to a future they had never imagined.

Standing with Dr. Iyengar in 2024 at JSU and GSU, fifty years after he first set foot on that campus, was nothing short of emotional. The world had changed, but in many ways, these institutions had not. HBCUs like JSU continue to face the same challenges: underfunding, lack of resources, and a system that still does not give them the support they desperately need. As we walked the halls, I saw the crumbling infrastructure, the outdated technology, the sheer disparity between these schools and their well-funded counterparts. But despite the hardships, there was resilience. The students and faculty at JSU, GSU, and other HBCUs continue to fight for their place in a world that too often overlooks them. Their determination mirrors that of Dr. Iyengar when he first began his journey—quiet, yet unwavering.

Image: Students at Grambling State University Presenting their Research work to Dr. S.S. Iyengar

Image: Students at Grambling State University with Dr. S.S. Iyengar, Dr. Bharat Rawal, Dr. Vasanth Iyer, Yashas Hariprasad and Dr. Sree Sanakam

Image: Dr. Vasanth Iyer and his mother (87 year old) hosting Dr. S.S. Iyengar, Veneeth Iyengar, Dr. Sree Sanakam and Yashas Hariprasad (September 26, 2024)

After his time at JSU, Dr. Iyengar's path led him to Louisiana State University (LSU), where he would serve for 32 years, including 20 as the Chairman of the Computer Science Department. Under his leadership, LSU's Computer Science program soared to national prominence, ranking in the top 30 of the National Research Council (NRC) rankings. But even as he reached these heights, he never forgot where he came from. He remained a fierce advocate for minority institutions, working tirelessly to secure funding, recruit top faculty, and provide opportunities for the students who needed them most. His journey at LSU was not just about personal achievement; it was about building bridges between majority institutions and HBCUs, ensuring that the gaps between them could be narrowed—if not closed entirely.

Image: Dr. S.S. Iyengar visiting Students at Jackson State University along with Dr. Natarajan Meghanathan, Veneeth Iyengar and Yashas Hariprasad

As we stood outside his old apartments in Jackson, where Dr. Iyengar had lived during his early years at JSU, a deep sense of nostalgia filled the air. The place was now in disrepair—run-down and a far cry from what it once was, much like the humble beginnings Dr. Iyengar faced. In those early days, he had little money, living in a poor and deteriorated area. Yet, as he spoke of those times, his voice carried a mixture of pride and humility. It wasn't about the conditions he endured or the accolades he would later earn; it was about the lives he touched along the way. It was about the students who, because someone believed in them, went on to accomplish extraordinary things.

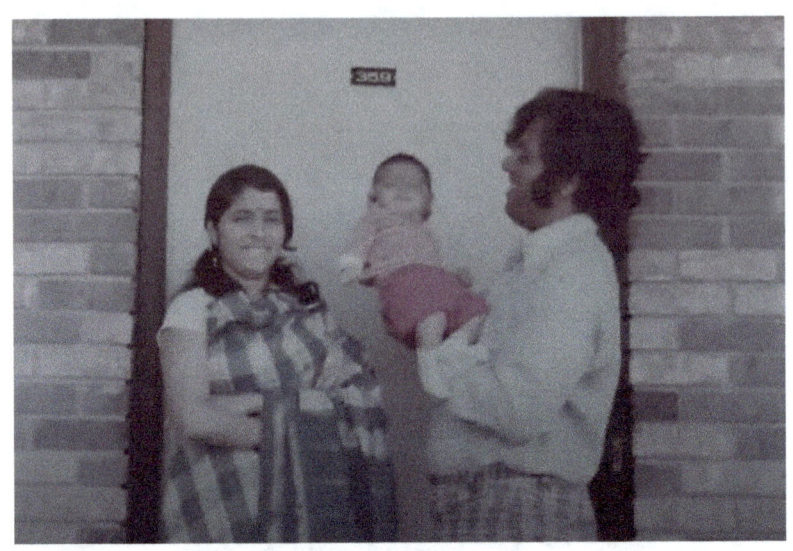

Image: **Dr. S.S. Iyengar and Manorama Iyengar with Puneeth Iyengar (5 Weeks old) in front of Camelot Apartments, Jackson, MS in the year 1975**

Image: in 2024, **Dr. S.S. Iyengar in front of Camelot Apartments, Jackson, MS**
This is where the story of Circuit Rider Began 5 decades ago

Later, we visited the small home he built (1400 Sq.Ft) in Baton Rouge (1518 Chippenham Drive), where he lived for 30 years after joining LSU. Unlike the modest apartment in Jackson, this home was a testament not to wealth or prestige, but to service. It stood as a place where ideas were born, nurtured, and shared with the world—a reflection of Dr. Iyengar's enduring commitment to uplifting others through education and mentorship.

Image: Dr. S.S. Iyengar in front of his house in Baton Rouge (2024)

But it wasn't just Dr. Iyengar's story that struck me; it was the ongoing struggle of the HBCUs we visited. These institutions, so rich in history and potential, are still fighting for basic resources in 2024. As I walked through JSU's aging classrooms and broken-down facilities, I couldn't help but feel a deep sense of urgency. How, in this modern age, could we still allow these universities to fall so far behind? Dr. Iyengar echoed my thoughts. He remains committed to supporting these institutions, and even today, he collaborates with JSU on a U.S. Army-funded Digital Forensic Center of Excellence grant. But his efforts alone are not enough. These schools need all the support they can get—from government, from society, and from us.

As we stood before the entrance of Jackson State University one final time, Dr. Iyengar reflected on how far he had come. Life, he said, is a cycle. He started his career at JSU, and now, fifty years later, he was returning as a collaborator on cutting-edge research. It was a full-circle moment—a poignant reminder that our beginnings often shape our journeys in ways we can't predict. His legacy is not just in the research he's done or the positions he's held, but in the lives he's transformed along the way.

Image: Dr. S.S. Iyengar addressing students at Grambling State University (September 2024)

There are several moments from our journey—pictures of Dr. Iyengar with students, faculty, and community leaders. In one, he stands proudly at the gates of JSU, a university that had transformed in the decades since his first arrival. In another, he poses with a group of students, many of whom credit his mentorship for their success. These images serve as powerful reminders of the countless lives he has touched—students who have gone on to break barriers, faculty who have become trailblazers in their fields, and institutions that continue to rise, despite the odds stacked against them.

This final chapter is not just a tribute to Dr. Iyengar's fifty-year legacy, but a call to action. His journey reminds us that the path to progress is not easy, but it is possible. It requires resilience, dedication, and an unshakable belief in the power of education. The seeds he planted in the fields of technology, scholarship, and minority education will continue to grow for generations to come. But it's up to us to carry that legacy forward.

Image: Dr. S.S. Iyengar and his Wife, Manorama Iyengar with their grandkids, Ranvir, Anjulie, and Vedh (Children of Dr. Puneeth Iyengar) – 2014 **

Image: Dr. S.S. Iyengar with his 6-month-old grandson Ramaraj (son of Veneeth Iyengar) – September 29, 2024

**

** "As Ramaraj, Ranvir, Anjulie, and Vedh grows, this moment will stand as a symbol of inspiration and legacy—a reminder of their

grandfather's unwavering dedication to education and service. One day, they may look back at this picture and draw strength from the life his grandfather led, inspiring them to carry forward the torch of service, impacting the lives of the people of Louisiana and beyond, just as Dr. Iyengar did." – Yashas Hariprasad

As I reflect on this journey, I am filled with both hope and determination. Dr. Iyengar's story is a testament to the power of one individual to create lasting change. He has proven that even the most daunting challenges can be overcome when we dedicate ourselves to the service of others. His legacy is not just his own; it belongs to all of us. It is a reminder that the work is not done, that there is still so much more to accomplish, and that each of us has the power to make a difference.

This epilogue does not mark the end of the story. It is, in fact, the beginning—an invitation to continue the work that Dr. Iyengar started fifty years ago. The circuit rider's journey is far from over. Now, it's our turn to ride.

Reflections on Five Decades of Progress: Dr. S.S. Iyengar's Journey in Advancing Education and Equity

Recently, I (Dr. S.S. Iyengar) revisited an evaluation of a program funded by the U.S. Army Research Office, where I have served as Director of a Center of Excellence that partners with key historically Black institutions, including Jackson State University, Grambling State University, and Florida A&M University (FAMU). Reviewing the outcomes of more than four decades of collaborative efforts, I felt an immense sense of fulfillment in seeing the progress achieved

---

in education, especially in the context of racial equity and opportunity.

The journey to educational equity, particularly for marginalized communities, is both arduous and inspiring. The strides made over these forty years stand as a tribute to the resilience and commitment of educators, students, and community leaders who've worked tirelessly toward this shared vision. HBCUs like Jackson State, Grambling State, and FAMU have been essential not only in delivering quality education but in fostering empowerment and a profound sense of belonging among their students.

Support from the U.S. Army Research Office has enabled us to craft innovative programs that enhance STEM education, enrich research opportunities, and prepare students of color for professional careers. This investment has gone beyond curriculum and faculty development; it has bridged academic and industry partnerships, opening doors for students to excel in competitive fields. We see the results clearly: rising graduation rates, increased participation in research, and more graduates advancing into postgraduate studies and successful careers.

Equally rewarding has been the community engagement that has grown alongside these educational initiatives. Partnering with local organizations, schools, and businesses has allowed us to create a holistic approach to education. Programs integrating service learning, mentorship, and internships enrich students' experiences and nurture a sense of civic duty that extends beyond academia.

The last four decades of progress are not just reflected in numbers; they are stories of lives transformed. Each graduation ceremony signifies not just academic success but dreams fulfilled and futures reshaped. Many alumni have become leaders, social justice advocates, and mentors for the next generation. Their successes remind us of the power of education to transcend race or socioeconomic barriers and redefine futures.

Yet, the path forward remains challenging. Systemic inequities continue to impact educational access and outcomes for students of color. We must maintain our vigilance, ensuring that resources are distributed equitably across institutions and that public and private sectors work hand-in-hand with academia to uphold and expand these gains.

As I reflect on this journey with Jackson State, Grambling State, and FAMU, I feel deeply hopeful. The future of education, especially for marginalized communities, is brighter than it was four decades ago. But our commitment must remain unwavering. Together, we must continue to dismantle barriers, uplift communities, and ensure that the transformative potential of education reaches all.

In revisiting this initiative, I am reminded that investing in education within communities of color is an investment in a more equitable and prosperous future. The progress we celebrate is a shared achievement, urging us to continue this work with persistence and purpose. Each step brings us closer to true educational equity, and with that, a future where opportunity is boundless and accessible to all.

-S.S. Iyengar

About the Author

Dr. Jerry F. Miller

Dr. Jerry F. Miller, Ph.D., is an assistant professor in the Department of Computing and Information Sciences at Florida Agricultural and Mechanical University. Prior to joining the Florida A&M faculty, Jerry served as Associate Director in several positions at Florida International University (FIU). He joined FIU's Applied Research Center in 2006 where he conducted research in renewable energy and global security. Later, he joined Dr. S.S. Iyengar in FIU's Knight Foundation School of Computing and Information Sciences. During this time, he continued his education, earning his Ph.D. in computer science and conducting research in computer network security, information security, and digital forensics with Professor Iyengar.

Before entering academia, Colonel Miller served in the United States Air Force as a rescue and special operations helicopter pilot, foreign area officer, and planning/programming/budgeting officer. He is an experienced leader and aviator with an extensive background in international human rights, as well as Latin American and Pacific Region foreign policy. He has lived and worked in many countries throughout the world but always finds his way home to the majestic live oaks of Florida.

Dr. Yashas Hariprasad

Dr. Yashas Hariprasad is a researcher at the U.S. Army-funded FINDS Digital Forensics Center of Excellence at Florida International University (FIU), where he pioneers advancements in artificial intelligence, cybersecurity, and digital forensics under the mentorship of the renowned Dr. S. S. Iyengar. His research focuses on developing cutting-edge AI algorithms for deepfake detection and quantum encryption frameworks, addressing pressing cybersecurity challenges in secure video transmission. He has authored and co-authored two books and more than 12 peer-reviewed journal and conference papers, published in esteemed platforms such as IEEE Transactions on Consumer Electronics, Elsevier, and Springer Nature. His work has garnered global recognition, featuring in major media outlets like Times of India –

Ahmedabad Mirror, ANI News, Press Trust of India, and The Tribune.

In addition to his research, Dr. Hariprasad has held significant leadership roles, including serving as the Co-founder of Puregem Naturals LLP, founding Vice President of the Artificial Intelligence and Coding Club at FIU, and as President of the Indian Student Association at FIU, where he championed initiatives promoting diversity, inclusivity, and collaboration. He has also mentored undergraduate and high school students through workshops and conferences, fostering the next generation of innovators.

His impactful contributions have earned him prestigious accolades, including the Medal of Appreciation from the Commander of the U.S. Department of Defense's Defense Information Systems Agency, the FIU Outstanding Scholar Award, and the FIU Best Graduate Student Award. Dr. Hariprasad earned his Ph.D. in Computer Science from FIU.

Ranvir Iyengar

Ranvir Iyengar is a sophomore at Chatham High School in Chatham, New Jersey. Over the years, Ranvir has demonstrated a significant interest in disparities that influence academic opportunities, medical outcomes, and economic inequalities. It is these interests that led Ranvir to do research into his grandfather, SS Iyengar, and his early years in the Deep South of the United States. In thinking about equity with respect to technologic exposure and social advancement in the deep South in the 1970's, Ranvir thought about how these opportunities have evolved for him and his generation, contributions he described in several chapters in the book Circuit Rider of Mississippi. These current day opportunities would not have been possible without the efforts of his grandfather and others of an earlier era.

Aside from already excelling in his course work, at even an early stage in his high school tenure, Ranvir has become a leader for his Mock Trial Team (regional finalists), Academic Bowl (winners of multiple competitions), Model UN, and many other extracurricular activities. He was the only freshman to play for the Chatham High School Varsity Tennis Team and continues to start for the team as a sophomore. As a freshman, Ranvir placed 2nd in a statewide essay competition about the legal system. He is already a co-author on at least 3 peer-reviewed manuscripts describing global and national health outcomes for a number of cancers and the inequities leading to disparities in outcomes. These articles will appear in the Annals of Surgical Oncology and The Prostate. Ranvir is leading several more manuscripts. Finally, Ranvir will be presenting some of this work at GU ASCO 2025, one of the few high school students presenting his work at the preeminent genitourinary cancer conference in the US. In the future, Ranvir plans to attend one of top US universities, focusing on the intersection of global economics, finance, and the law.

www.ingramcontent.com/pod-product-compliance
Lightning Source LLC
Chambersburg PA
CBHW061741120626
46550CB00005B/1846